5/2015

D0609810

Experimental Evolution

AND

The Nature of Biodiversity

ℭℬ

REES KASSEN

Department of Biology

University of Ottawa

ROBERTS AND COMPANY

Greenwood Village, Colorado

Experimental Evolution and The Nature of Biodiversity

Roberts and Company Publishers, Inc.
4950 South Yosemite Street, F2 #197
Greenwood Village, CO 80111 USA
Tel: (303) 221-3325
Fax: (303) 221-3326
Web: www.roberts-publishers.com
Email: info@roberts-publishers.com

Publisher: Ben Roberts
Production management: Kathi Townes at TECHarts
Artists: Lineworks
Copyeditor: Chris Thillen
Proofreader: Kathi Townes
Composition: Kristina Elliott at TECHarts

Front cover:© 2013 Artists Rights Society (ARS), New York/ADAGP, Paris

ISBN: 978-1-936221-46-2

Library of Congress Cataloging-in-Publication Data
Kassen, Rees, author.
 Experimental evolution and the nature of biodiversity / Rees Kassen, Department of Biology, University of Ottawa.
   pages cm
   Summary: "Why and how did life become so diverse? This has been a central problem in biology. Experimental Evolution and the Nature of Biodiversity explores how diversity evolves in microbial populations that occupy some of the simplest environments imaginable--laboratory test tubes"-- Provided by publisher.
 Includes bibliographical references and index.
  ISBN 978-1-936221-46-2 (paperback)
 1. Biodiversity.   2. Evolution.   3. Biology, Experimental.   I. Title.
 QH541.15.B56K37 2014
 576.8--dc23
                              2014009628

10 9 8 7 6 5 4 3 2 1

To Sascha and Noah, les chevaliers de Laudou et Ossington,
and to Genevieve, notre reine

Contents

Preface

This is a book about biodiversity. It is also a book about experimental evolution. The book explores what experimental evolution has told us about how biodiversity evolves.

A synthetic treatment of the evolution of biodiversity has not been readily available. Biodiversity studies have for the most part been loosely organized into camps focused on particular levels of biological organization. Population geneticists concerned themselves with genetic variation within populations, community ecologists with species diversity in communities, and paleontologists with patterns of diversity in the fossil record. On a more practical level, conservation biology focused on the various threats and insults faced by biodiversity from human activities and how these might be mitigated. All these approaches share the same fundamental concern: how ecologically very similar yet phenotypically distinct types coexist in natural communities.

In other words, the study of biodiversity focuses almost exclusively on *maintenance* but has largely ignored *origins*. When origins are considered, the explanations provided are often broad and general. Natural selection is commonly cited as a major factor driving diversification, but the details of how it generates diversity are usually not made explicit. Too often, researchers simply assume that natural selection had been powerful enough in the past to generate the "right" sorts of adaptations and trade-offs necessary for diversity to be maintained in the present. Whether or not this is actually the case is rarely examined in close enough detail for anyone to judge.

Such glossing over the details of how biodiversity evolved could be excused because of a natural bias in what most evolutionary biologists choose to study: large, multicellular organisms with complex life cycles and long generation times. To study the evolutionary processes in general, and biodiversity in particular, requires that we redirect our attention to very small and, to some, less sexy organisms such as microbes. Microbes reproduce in minutes, not years, so it is possible to conduct evolutionary experiments involving hundreds or even thousands of generations by keeping populations of microbes growing continuously in the laboratory. Under these conditions, genetic variation naturally arises through mutation, and this leads inevitably to natural selection. This is experimental evolution, a technique that allows us to follow the emergence and fate of genetic variation directly, and in the process to test evolutionary theories about the origins of biodiversity.

The past 15 to 20 years have seen a veritable explosion of studies using experimental evolution to explore the origins and fate of diversity. This book summarizes and synthesizes what these studies can teach us about the evolution of biodiversity. Its aim is to provide a more explicit account of how biodiversity evolves than has been available up to now and, in the process, provide some insight into the fundamental properties and character—the nature—of biodiversity.

Experimental evolution is founded on the assertion that the principles governing adaptation and diversification in a microcosm are the same as those in natural systems. In other words, it takes as its starting point that what is true for microbes in test tubes is equally true for all other living organisms. A large community of researchers in biodiversity are not yet comfortable with this assumption, and so they have either ignored experimental evolution altogether or remained highly skeptical that it says anything meaningful about the processes supporting diversity in the real world. This skepticism is understandable but misguided. Microbial experimental evolution can and should play an important role in the broader effort to understand biodiversity, most importantly because it affords the opportunity to strip away the complexity of nature so that we can focus on the fundamental principles that govern the evolution of all life. My hope is that this book will go some way toward convincing the skeptics that experimental evolution is not only valid but essential part of the arsenal of techniques for understanding the diversity of life.

Recognizing that not everyone feels the same way, I have put forth special effort in this book to make the experimental evolution research program more accessible to a broad readership. First, I have devoted Chapter 1 to the history and basic techniques used in experimental evolution, as well as a discussion of some of the paradigmatic experimental systems used. Second, I point out, wherever appropriate, the limits to inference of the experimental evolution enterprise, motivated by the conviction that we should at least be honest about what these experiments can and cannot say about how diversity evolves in the real world. Third, I begin each chapter with a short narrative of a real-world example that introduces the main evolutionary problem that experimental evolution helps answer. My hope is that these vignettes serve to ground the book in the sorts of systems most biologists are familiar with and, through the lens they provide, afford us a sharper picture of where experimental evolution can help push the field forward, and where its limitations lie. Some readers may find the vignettes more distracting than useful. If so, not much is lost by skipping them. Nevertheless they have been fun to research and to write, so I do hope you find them interesting, and hopefully valuable, to read.

Acknowledgments

Although this book has just one author, any project of this size is a team effort. I want to thank Ben Roberts and his team at Roberts and Company for their help in seeing this book through to completion. Kathi Townes, who managed the publication process, deserves special mention for not getting too frustrated with delays and disorganization on my end. Thanks also to Emiko Paul, the cover designer, and Chris Thillen, the copyeditor, for your creativity and help along the way. Ben himself was an outstanding and ever-patient supporter of and believer in this project from the start. Thank you, Ben, for everything.

I also want to thank the many people who I have talked to about the content of the book and who provided comments and feedback along the way. Mike Whitlock, especially, encouraged me to write the book and has reviewed the entire thing multiple times along the way to completion. Others have provided comments on specific chapters or ideas; they include David Reznick, Thomas Bataillon, Ophelie Ronce, David Currie, Luke Harmon, Bill Hanage, and Mark Vellend. A special thanks goes to Ophelie and her colleagues in Montpellier for giving me a welcoming and collaborative space in which to write the final draft. The book has benefited immensely from conversations during that time with Guillaume Martin, Jean-Nicolas Jasmin, Thomas Lenormand, and Oli Kaltz. It has also benefited from discussions with colleagues and friends at the University of Ottawa: Howard Rundle, Scott Findlay, David Currie, and Risa Sargent. My lab group has been an important source of inspiration and a sounding board for many of these ideas. Thank you to the current group: Susan Bailey, Anita Melnyk, Oleksiy Teselkin, Sabrina Slade, Elizabeth O'Reilly, Jeremy Dettman, Aaron Hinz, Susanne Kraemer, Gabriel Perron, and Nicolas Rode, as well as past members for being patient with me and for pushing me to think harder about what this is all for. Each of you has had more influence on this book than you might realize. I have written the book in part as the sort of text I would have wanted when I was starting graduate school. I hope you find at least some of what is in here useful as you move on in your careers. It goes without saying, though it still needs to be said, that any mistakes or omissions in the text remain mine alone.

A special note of thanks to Karen McCoy and Thierry Boulinier, who made their own contribution by agreeing to exchange houses—and in many ways, lives—during the last six months of writing this book while my family and I were on sabbatical in the south of France. Having the beautiful and quiet space of Laudou was a true gift. It gave me the space and time to finish the manuscript for the book, and it gave my family the opportunity to grow in new and unexpected ways. I hope your experience in Ottawa allowed you and your family to do a little growing in similar ways. Thank you.

Finally, thank you to my family for their incredible support over the past few years while I was putting together this book. Genevieve, you have believed in me ever since day one. You have created time for me where it otherwise wouldn't have been, looking after our children and our home so that I could focus on this project. Thank you for that and for your love and friendship every step of the way. I look forward to doing the same for you when the time comes. My children, Sascha and Noah, were inspirations to me throughout. Thank you, boys, for understanding that Daddy had work to do sometimes. I'll be joining you soon to watch the snail races. My parents, Barry and Susie, and my sister, Robyne, have always supported me, even if they profess not to understand what this is all about. Thank you for your patience and trust. Mom and Dad, you sparked an early interest for me in all aspects of diversity, natural and otherwise. Thank you.

REES KASSEN
Laudou, France, and Ottawa, Canada
October 2013

Introduction

On an evening in July 1858, Charles Darwin and Alfred Russel Wallace's joint paper on evolution by natural selection was read at a meeting of the Linnean Society. Soon after, it was published in the society's journal. By all accounts, the response to one of the most important ideas of the nineteenth century was rather subdued. Neither author was present when the paper was read, and little discussion took place afterward. Indeed, so lifeless were the proceedings that the society president, Thomas Bell, remarked in his annual address that the year just past had been less than inspirational, containing no revolutionary discoveries.

That's how it often goes with profoundly new ideas. After years in gestation, they are born into a world that greets them with a decidedly passive and uninterested "uh-huh." Did the members present think the idea was so obvious as to be self-evident? Did they need more time to appreciate its importance, or were they simply bored? Because so few accounts of the event were written, we may never know the answer. Whatever the case, this was the moment, lackluster though it seems, that introduced the world to natural selection—the foundational principle for understanding adaptation and the origins of life's diversity.

Why and how did life become so diverse? This has been the central question—or more accurately, the obsession—in biology. A history of answers would be a history of biology itself, if not a fairly accurate barometer of the attitudes and mores of society at large. It's also the question that motivates this book. Darwin and Wallace's idea seemed, in one fell swoop, to provide the answer. Natural selection leads to adaptation, causing an often exquisite "fit" between organism and environment. Repeated again and again, over time and in different environments, this process inevitably leads to the emergence of diverse populations and eventually to species.

Yet our understanding of the factors governing the process of diversification has not gone much beyond Darwin and Wallace's original formulation. The bulk of existing theory on diversity in both population genetics and ecology concerns the maintenance of diversity, but not its origins. Paleontologists, on the other hand, have traditionally focused on broad-scale patterns in the evolution of Earth's diversity—at the cost of ignoring the more detailed mechanisms governing how adaptation leads to lineage splitting and eventually diversification.

It's interesting to consider why so little progress has been made. George Williams may have been essentially correct when he wrote in his conclusion to *Adaptation and Natural Selection* (1966) that nothing was ever really demanded of the theory. Darwin and Wallace had produced what was by all accounts an adequate and complete explanation. Given that natural selection generated adaptation and, eventually, diversity, the focus could shift to asking more pointed questions. These might include whether a trait is adaptive, how genetic variation (or species diversity) can be maintained in the face of natural selection, and the large-scale drivers of diversity at continental levels or through geologic time.

Another reason for the lack of progress is that evolutionary biology has been largely an inferential rather than experimental science. This is, of course, necessarily so in paleontology, where fossils are all we have to study. But it has been equally true for those studying contemporary populations, because most researchers study large multicellular organisms. This does not mean that experiments are absent, or that nobody has advocated studying the process rather than just the results of evolution (see, e.g., Dallinger 1887; de Varigny 1892; Silvertown et al. 2006), but rather that the emphasis has been mostly on interpreting process from pattern. If all we have to study are large, long-lived organisms, then naturally the scope for doing truly evolutionary experiments—those where the dynamics of genes or gene frequencies are tracked alongside measures of fitness and diversity—is extremely limited.

The intense focus on speciation has not helped matters, either. The speciation research program has been outstandingly successful in emphasizing the special problem that speciation poses for diversification: how reproductive isolation evolves in the face of gene flow (Coyne and Orr 2004; Schluter 2000a). The concept of ecological speciation, where reproductive isolation develops alongside or as a by-product of adaptation to different environments, has emerged as perhaps the leading explanation for the evolution of reproductive isolation. However, this emphasis on reproductive isolation has two unfortunate consequences for understanding how diversity evolves. First, the evolution of reproductive isolation is taken to be the starting place, rather than the endpoint, of diversification. As a result, local adaptation is often (though by no means always) assumed rather than demonstrated in studies of speciation. Second, a focus on reproductive isolation necessarily restricts attention to sexually reproducing species and ignores the much larger, though often less conspicuous, world of asexual species. Ecological divergence is often taken to be the defining characteristic of asexual species, especially in microbes. The emphasis on reproductive isolation has thus eclipsed the broader, more fundamental problem of understanding how ecological diversity evolved in the first place.

The end result is that our explanations for the origins of biodiversity often fail to provide useful insight. They are either too general, as when one states that natural selection causes biodiversity, or too specific, as when the focus is exclusively on the evolution of reproductive isolation. Neither explanation takes us very far toward a theory of diversification. Such a theory should be capable of providing insight into the mechanisms causing diversification: what are the sources of natural selection leading to diversification, and under what conditions do they operate? Do some environments support more diversity than others and, if so, why? What happens to diversity in the long run? Is it eventually lost, or does it persist in perpetuity? Does the presence of diversity lead to ever more diversification, or are there limits to the amount of diversity that a community can support?

This book is an attempt to provide some answers to these questions. It offers an account of how biodiversity evolves in some of the simplest biological systems studied to date, microbial populations occupying the defined environments of laboratory test tubes. These experiments are far removed from the natural world that brings most of us into the study of evolutionary biology and natural history to begin with. But they're sufficient representations of that world that we can begin to develop a more complete and comprehensive account of the ecological and genetic factors responsible for diversification. Experimental evolution with microbes thus opens a window onto the very nature of biodiversity—its properties, its character—that is impossible to attain in most study systems.

This book presents an explanation for how ecological diversity—the origin of genetic variation in fitness among types that do, or may, compete for similar resources—evolves. Doing so requires breaking down the process of evolutionary diversification into its component parts. In essence, there are just two. The first is adaptation, the core process driving adaptive differentiation. The second is the process of lineage splitting itself that gives us two (or more) recognizably distinct populations or species. For the most part, the connection between adaptation and diversification has not been studied in depth, although it's widely recognized that adaptation is likely to be an important driver of diversification. The major goal of this book is to make the connection between adaptation and diversification more explicit, and by doing so begin to sketch a more comprehensive theory of diversification.

To do this, the book draws on the growing literature of microbial experimental evolution. The great advantage of using microbial systems is, of course, that a theory can be tested directly by conducting evolutionary experiments that follow the emergence and fate of genetic variation, under conditions that the experimenter controls. Combining the results of these experiments with cost-effective whole-genome sequencing technologies makes it possible, for the first time, to gain a truly comprehensive view of the genomic and ecological factors governing adaptive diversification.

The inferential strength of these experiments is potentially strong, but must not be oversold. Microbial evolution experiments typically involve very large population sizes and fairly long durations, between hundreds and tens of thousands of generations. They also often start from a single clone and so involve adaptations that are constructed from mutation rather than standing variation. The results of these experiments thus bear most directly on natural systems that are comparable in these same respects. It's probably safe to say that this will be true of some microbial species, with the caveat that the environments studied in the laboratory are no doubt far less complex than those encountered in nature. The results are less directly comparable to other natural systems where organisms are larger and effective population sizes smaller. The results of the work presented here should thus be interpreted in the appropriate context: as examples of how the evolutionary process plays out in (usually, though not exclusively) relatively simple environments when population sizes are large and adaptation is fueled by mutation.

HOW THIS BOOK IS ORGANIZED

This book has two parts. The first concerns adaptation, the second diversification. The material reviewed comes from the literature on microbial experimental evolution. The experimental techniques are derived primarily from microbiology and so involve some principles that may not be familiar to many readers. Chapter 1 gives a brief account of the basic theory and methodology used in these experiments. It also briefly describes some canonical experiments in the field, in particular Richard Lenski's long-term selection experiment using *Escherichia coli* and the model adaptive radiation by *Pseudomonas fluorescens* first described by Rainey and Travisano (1998). The chapter also discusses the basic principles behind the use of bacteriophage in experimental evolution. Examples from these experiments recur throughout the book.

The first part (Chapters 2 through 5) is about adaptation to successively more complicated ecological scenarios. Chapter 2 considers adaptation to a single, uniform environment and introduces recent approaches to developing a quantitative theory of adaptation. Key questions to be considered here include the number of genes involved in adaptation, their fitness effects, and the repeatability of the evolutionary process. Chapter 3 discusses adaptation to divergent environments and the ease with which trade-offs among environments evolve. The evolution of diversity hinges on the emergence of such trade-offs, yet a predictive framework for understanding how trade-offs evolve has been lacking. Moreover, experimental studies have had a harder time than expected demonstrating trade-offs. This chapter describes some first steps taken toward understanding the evolution of trade-offs. Chapter 4 addresses the problem of selection in environments that vary in space and time. In principle, the niche breadth of a lineage—the range of conditions across which fitness can be maintained—should evolve to match the extent of environmental variation. Under this view, generalists and specialists should evolve under more variable and less variable environments, respectively. However, much insight into the evolution and maintenance of diversity comes from understanding the exceptions to this rule, and these exceptions are the focus of this chapter. Chapter 5 turns attention to the molecular genetics of adaptation. Are there regular rules governing the kinds of genes or genetic systems involved in adaptation, at least in the short run of selection experiments lasting hundreds to thousands of generations? What genes are typically involved?

Diversification is the subject of the second part of the book (Chapters 6 through 9). The simplest way to understand the evolution of ecological diversity is as a natural extension of the process of adaptation. Yet a persistent train of thought in the literature sees diversification as something apart from the normal, secular process of adaptation. Nowhere is this latter view more apparent than in discussions of adaptive radiation, a field that has attracted at times somewhat extravagant explanations for the apparently rapid and spectacular diversity that its name suggests. These two opposing views have never been adequately reconciled, and doing so is the main goal of this section.

In this part, we first consider the factors that control three essential properties of a diversification event: the extent of phenotypic divergence among derived types (Chapter 6) and the number and rate at which new, distinct types evolve (Chapter 7). The aim here is to ask whether diversification is a genuinely unusual or distinct process that cannot be understood as the consequence of adaptive differentiation to distinct conditions. The short answer is no, but certain properties of diversification events—their speed or their extent—can make some instances of diversification stand out against the background of a more secular process of adaptation and differentiation. Paramount among these properties is adaptive radiation, the unusually rapid diversification of niche specialists. Because adaptive radiations hold such a revered place in the minds of many evolutionary biologists, this book devotes an entire chapter to them (Chapter 8). Chapter 9 considers the molecular genetics of diversification. Are some genes or genetic systems prone to diversification? Surprisingly, perhaps, the answer to this question turns out to be a tentative yes.

Chapter 10 sums up the key points of the book in an attempt to provide a preliminary sketch for a general theory of diversification. Any such theory must include an account of both the ecological and genetic mechanisms involved in diversification. The great advantage of the microbial experimental evolution approach is that both ecology and genetics can be studied in the same experiment. On the other hand, many researchers remain somewhat skeptical of the value of experimental evolution for gaining insights into how diversification occurs in more natural systems. Chapter 10 includes a number of responses to the skeptic's position. Finally, it offers some thoughts on the most compelling and exciting directions forward for the science of experimental evolution and what it can tell us about the nature of biodiversity.

 CƷ

Adaptation

This part of the book aims to address how far we have come, since that evening in July 1858, toward understanding adaptation and how it contributes to the evolution of diversity. The focus here is on the process of adaptive evolution and the properties of the genes involved.

This is quite a distinct approach from the one that dominated discussions in the literature on adaptation during the latter half of the twentieth century. The emphasis then was on trying to decide whether a trait or behavior was adaptive. The result was the development of a sophisticated comparative method, optimization theory, and vigorous research programs in ecology, physiology, and behavior that sought to demonstrate the adaptive utility of organismal characteristics. This approach to studying adaptation put us in the position of knowing quite a lot about adaptations themselves, but very little about how they are constructed.

To understand how adaptations are built, we need to know something about their raw materials. These are beneficial mutations, genetic changes that increase fitness relative to the ancestral genotype they are derived from. Adaptations are constructed through successive substitution of beneficial mutations, driven by natural selection. The challenge is that beneficial mutations are typically so rare that they have been difficult to study. Deleterious mutations are far more abundant and thus much easier to study. Indeed, because they are so frequent, deleterious mutations are the bread and butter of molecular and developmental genetics. Their large and sometimes grotesque effects on phenotype are used for making inferences about the genetic architecture of metabolism, physiology, and development. This knowledge, although valuable in its own right, tells us little about the "stuff" of adaptive evolution: beneficial mutations.

This situation is starting to change. In the last 15 years or so, there has been a resurgence of interest in models of adaptation, and some key predictions of these models have been empirically tested. The central questions concern the magnitude of changes wrought by natural selection when it acts on genetic variation: Is most adaptation slow and gradual, being driven by very small changes in phenotype? Or, is it rapid and steplike, fueled by mutations of large effect? These questions have been with us, in some form or another, throughout the history of evolutionary biology. To them, we need to add others in order to understand more fully the process of adaptive diversification: What are the consequences of adaptation to one set of conditions for fitness under different conditions? What factors—ecological or genetic—speed up or slow down the process of adaptive differentiation? How many genes are typically involved in adaptation, and what do they do? In this part, we'll start answering some of these questions.

CHAPTER I

Experimental Evolution: An Introduction

The idea of using microbes to test theory in evolutionary biology is almost as old as the principle of natural selection itself. The first published account was by a Wesleyan clergyman named William Henry Dallinger (1839–1909; **Figure 1.1**), who at the time was president of the Royal

(a)   (b)

Figure 1.1 (a) The good Reverend W. H. Dallinger, the first person to use microbes for tests of evolutionary theory. *Source:* Wellcome Library, London. (b) Dallinger's custom-built incubator. (Reproduced with permission from W. H. Dallinger. The President's Address. *J. Royal Microscop. Soc.,* 1887 published by Wiley and Sons).

Microscopal Society in the United Kingdom. In his president's address to the Society for 1887, Dallinger asked "whether it was possible by change of environment, in minute life-forms, whose life-cycle was relatively soon completed, to superinduce changes of an adaptive character, if the observations extended over a sufficiently long period" (Dallinger 1887, 191).

Dallinger started his research with three morphologically distinct flagellated unicellular eukaryotes, what he termed "monads," and allowed them to grow in a purpose-built incubator. Over the next seven years, he gradually increased the incubator temperature. The cell populations grew slowly and episodically. Dallinger found that he could raise the temperature a little, but then he had to let the cultures stay at that temperature for some time, often weeks, to ensure the populations would not crash. He made detailed, meticulous drawings of the cells throughout the experiment, noting that no morphological changes occurred. Eventually he managed to get the populations to grow at 158°F, well beyond their normal thermal tolerance limit of 140°F. The experiment was prematurely terminated because the incubator collapsed, for reasons that were not reported. Just before the accident, though, Dallinger returned the populations to their ancestral temperature of 60°F. None grew. He reasoned that the populations had adapted to the increasing temperatures by natural selection and in so doing had lost the ability to grow at their normal temperature.

This experiment was the very first direct test of Darwin's theory of natural selection. Not the first demonstration of selection, mind you. Animal breeders were well aware that they could select for certain traits by carefully choosing parents with the desired characteristics. Dallinger's work was different, though. He did not follow the animal breeder's lead by choosing parents with higher thermal tolerance. Rather, he just slowly increased the temperature and let nature do the rest.

The result, of course, was adaptation by natural selection. In the process, he provided the first example of an evolved trade-off, in this case between growth at high and low temperature. He was also probably the first to demonstrate the power of genetic mutation in causing adaptation, though he probably didn't know it at the time (the rediscovery of Gregor Mendel's work on inheritance in peas was still over a decade away). With the benefit of hindsight, though, it is tempting to see the resting periods he gave his populations before each incremental increase in temperature as the time required for the appearance and substitution of mutations allowing growth at ever-higher temperatures.

Beyond these valuable insights into the process of adaptation, Dallinger's experiment made an additional contribution to the field of evolutionary biology: It introduced the idea that evolution can be studied on timescales that are amenable to experimentation and analysis. After all, the rapid generation times and large population sizes of microbial populations are ideal conditions for natural selection. With relatively little additional development, the approach Dallinger took more than a century ago remains in use today and is a powerful method for studying some of the most fundamental questions in evolutionary biology.

Interestingly, Darwin knew of Dallinger's work because Dallinger had written to him about it. In a response dated July 2, 1878, Darwin wrote: "I did not know that you were attending to the mutation of lower organisms under changed conditions of life; and your results, I have no doubt, will be extremely curious and valuable. The fact which you mention about their being adapted to certain temperatures, but becoming gradually accustomed to much higher ones, is very remarkable. It explains the existence of algae in hot springs. How extremely interesting an examination under high powers on the spot, of the mud of such springs would be" (Dallinger 1887, 191–92). Darwin appears much more excited by the opportunity microscopes offered to observe algae in hot springs than by the opportunity to test his theory of natural selection. Perhaps Darwin valued experiments more for the facts they contributed to the store of knowledge than as a means of testing hypotheses.

Since Dallinger's time, microbial selection experiments have been used occasionally in ecology and evolution, often to good effect. Gause (1934) introduced the competitive exclusion principle through experiments with two species of *Paramecium*. He showed that although each species grew well on its own, when placed in the same vessel, one inevitably displaced the other. Tilman (1977; 1982) extended this work by showing that two species of algae could coexist if each species was limited by a different resource. Atwood, Schneider, and Ryan (1951) and Novick and Szilard (1950) introduced the idea of periodic selection by showing that in long-term cultures of *Escherichia coli*, a regular replacement of genotypes occurred that was associated with adaptation. And, of course, the evolution of drug resistance in pathogenic bacteria and fungi continues to provide one of the most compelling and important examples of adaptive evolution in action.

The history of molecular biology can be mined for other examples of adaptation and diversification in laboratory culture. In the early twentieth century, Felix D'Herelle, a codiscoverer of bacteriophages—viruses that infect bacteria—showed how repeated passage of phages could lead to increased virulence and changes in the specificity of infection (Summers 1999). Many microbiologists, moreover, recognize that a "well-behaved" laboratory culture should grow vigorously in the broth of a growth medium and should not form colonies of variable morphology or stick to the walls of glassware. Researchers can spend many weeks, and sometimes months, domesticating wild strains of bacteria for laboratory use. They selectively subculture only those cells that grow in the broth phase and do not generate variation spontaneously. It is easy to see that such domestication probably involves adaptation to laboratory culture, although microbiologists do not tend to use the term *adaptation* to describe this process (they prefer *acclimation*).

Despite these examples, throughout most of the development of ecology and evolution in the twentieth century, microorganisms remained mostly at the margins of mainstream studies of adaptation and diversity. Only in the last 15 to 20 years, in evolutionary biology especially, has there been a concerted effort to focus attention on microbial experimental evolution as a means of testing fundamental theories about adaptation and diversity. Let's start the chapter by looking at how this work is done.

METHODS IN EXPERIMENTAL EVOLUTION

The logic behind all selection experiments is straightforward. Variation that arises through mutation or preexists in the population generates competition among genotypes as the population grows and uses up available resources. The type that is fittest, in the sense of being the fastest replicator under the prevailing conditions of growth, comes to dominate the population through natural selection. This leads to a steady replacement of one genotype by another and so to a change in the genetic makeup of the population. Consequently, evolved populations will outcompete the ancestor in the environment in which they have been selected. In ecology, it is common to start with a highly diverse population (or collection of species) and follow the fate of diversity over the course of a few tens of generations (McGrady-Steed and Morin 2000; Sommer 1985). Evolutionary experiments typically begin with a single strain and last for hundreds or thousands of generations, during which new variation arises through mutation. The focus in this book is on evolutionary experiments.

The techniques for studying adaptation and diversity in the laboratory have not changed much from Dallinger's time. The basic equipment used to propagate microbial populations can be found in any standard microbiology laboratory: test tubes, pipettes, an autoclave for sterilization, incubators, and the chemicals needed to make nutrient media. The analytical tools are borrowed directly from quantitative genetics, agronomy, and animal breeding (Bell 2008; Elena and Lenski 1997; Falconer 1981; Lynch and Walsh 1997). The primary aim of these analyses is to evaluate how much a population has adapted relative to a common ancestor. This process requires separating genetic from environmental effects. Therefore, assays invariably involve some form of common-garden experimental design, different genotypes being raised in the same environment. Preferably, the environment used in a common-garden experiment is the same as that under which the population evolved during the selection experiment, as only then can one be confident that whatever changes are observed are solely due to genetic causes. Because most studies are conducted with asexual haploid organisms, estimates of the genetic contribution to adaptation do not require complicated breeding designs. Researchers simply estimate the fitness of the evolved population and the ancestor it was derived from, and then compare the two. The difference between the two fitness estimates is a measure of the amount of adaptation that has occurred during the experiment.

Population Growth

Evolution experiments require a population to grow in perpetuity, or for at least as long as the theory requires and the experimenter has patience. Two techniques are commonly used in microbial selection experiments: batch culture and continuous culture.

Batch culture involves inoculating a small volume of cells into liquid medium in a test tube or flask, letting the population expand for a defined period, usually 24 hours, and then transferring a sample of cells (called an aliquot) into fresh medium. Occasionally, agar plates are used, and in that case the colonies are washed or picked off the plate at regular intervals and seeded onto a new plate.

Continuous cultures maintain populations growing at a constant rate (chemostat) or density (turbidostat) by a constant inflow of fresh medium and removal of spent medium that contains both cells and waste. Turbidostats are not extensively used in experimental evolution, though perhaps they should be, because this system closely approximates a Fisher-Wright-Moran population model of constant, finite size.

Population growth in batch culture or on plates typically follows an S-shaped pattern that resembles the logistic model familiar from elementary ecology and population biology (**Figure 1.2a**). In the logistic model, population growth rate is given by the expression

$$\frac{dN}{dt} = rN\left(1 - \frac{N}{K}\right)$$

where N is the population size, r is the intrinsic rate of growth, and K is the carrying capacity of the environment (the maximum number of individuals supported). Population size at time t (N_t) is found from the solution of the previous equation:

$$N_t = \frac{KN_0e^{rt}}{K + N_0(e^{rt} - 1)}$$

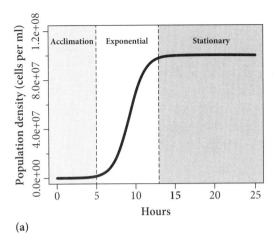

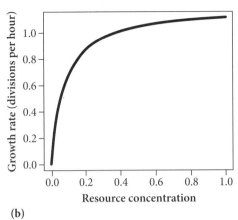

(a)   **(b)**

Figure 1.2 (a) Logistic growth of a microbial population in batch culture, showing the three phases of growth commonly recognized by microbiologists. (b) Growth rate as a function of resource concentration, after Monod (1949). In continuous cultures, the growth rate is set by the dilution rate of the culture vessel.

Microbiologists recognize three growth phases in batch culture: an acclimation phase where there is little or no population growth while individuals adjust physiologically to the medium, an exponential phase during which population growth rates are near their maximum and resources are in excess, and a stationary phase where population density is maximal, resources have been largely used up, and growth appears to have stopped.

The logistic model is appropriate for most microbial species because it provides a straightforward way of capturing the dynamics of growth during times of resource abundance and scarcity. It is only one of many equations, and it may not be appropriate for all purposes. Examples of alternative approaches can be found in Gifford and Schoustra (2013), who study population growth in filamentous fungi growing on agar plates, or during serial transfer of phage on an excess of host cells (Bull et al. 2011).

Continuous culture populations are kept in a state of perpetual resource limitation. This procedure ensures that the medium flows through the culture vessel at a rate lower than the maximal growth rate and prevents the population from being washed out. The population growth rate, μ, is modeled as a function of the concentration of a limiting resource, R, using a hyperbolic equation first suggested by Monod (1949; **Figure 1.2b**):

$$\mu = \mu_{max} \frac{R}{S_R + R}$$

where μ_{max} is the maximum growth rate, and S_R is the concentration of resource when growth rate is half maximal, $\mu_{max}/2$. If we assume that each unit of resource yields a constant fraction of individuals, then the density of the population in the culture vessel at equilibrium depends on the yield coefficient and the concentration of resources used by the population. It is thus possible to maintain a culture of constant density, so long as the yield coefficient doesn't evolve. Further details on the equations used to model chemostat growth can be found in Dykhuizen (1990).

The Number of Generations

The number of generations is estimated by the dilution rate for both batch and continuous culture. This value is straightforward to calculate for a chemostat, because the dilution rate sets the growth rate. The number of generations is just the dilution rate multiplied by the duration of the experiment. In batch culture, the dilution rate depends on the volume of cells transferred and the volume of the growth medium. When a small volume, such as 0.1 milliliter (mL), of cells is inoculated into 9.9 mL of fresh medium, the population must increase by a hundredfold to prevent being washed out. We can therefore calculate the number of doublings that must have occurred as $\log_2(100)$, or 6.64 generations. This is a minimum estimate of the number of generations in batch culture. Most bacterial species used in experimental evolution reach stationary phase within the first 8 to 12 hours, and so populations will spend at least half their time in stationary phase before being transferred if they are on a daily transfer regime. The population is not likely to be completely quiescent during this time, and a number of experiments have shown that adaptation to stationary phase conditions does occur (Meyer et al. 2011; Zambrano et al. 1993; Zinser and Kolter 1999). These results imply that substantial population turnover must take place even in times of resource scarcity. The number of cell divisions that occur during stationary phase is not a topic that has been extensively investigated.

Measuring Fitness

The easy part in a selection experiment is actually doing the selection itself, because this involves simply maintaining a growing population. Nature, by introducing mutations that create the opportunity for natural selection, does the rest. Assaying the effectiveness of selection in generating adaptation and supporting diversity is where the real work happens. Adaptation occurs when the fitness—that is, the replication rate—of the evolved strains or population is greater than that of the ancestor when measured in the same environment where selection has occurred. To estimate the quantity of diversity in a population, researchers must isolate individual genotypes from the population and estimate the variance in fitness among them. Occasionally, diversity can be estimated through phenotypic measures such as colony morphology or cell size; but these, like all phenotypic measurements, must be treated as traits that are correlated with fitness.

Fitness itself can be measured in either of two ways. The first estimates growth parameters (of either a genetically mixed population or a single, genetically unique strain) such as the acclimation period, maximum rate of growth, and carrying capacity or stationary phase density in the absence of any competitors. The second approach involves mixing the evolved strain or population with a common competitor, usually the ancestral strain from which the evolved populations were derived, and measuring the rate at which one excludes the other. This method is preferable because it is an integrated measure involving all phases of the growth cycle, and, when the competitor is the ancestor from which the lineages have evolved, provides a direct estimate of how much adaptation has occurred over the course of the experiment. A variety of neutral genetic markers with striking phenotypic effects, often on colony morphology or color, can be used to distinguish evolved and competitor strains.

The ability to freeze and revive strains makes it possible to obtain a complete fossil record of the experiment. It also makes it easy to estimate changes in fitness throughout the experiment by comparing the evolved lineages directly against their common ancestor. Estimates of absolute fitness for a genotype or population grown in isolation are usually given either as population growth rates, r, or as stationary phase density (aka carrying capacity, K). This is done either by fitting an appropriate growth model, like the logistic model, to data or by plotting log-transformed density over time. When r is measured on a logarithmic scale, the difference in growth rates between two strains (or populations) is a measure of the selection rate constant (or selection coefficient), s, between them. The selection coefficient is a measure of how rapidly one strain would outcompete the other in competition.

Estimates of fitness based on growth parameters alone may not capture all the relevant features of competition, so it is often preferable to measure fitness by mixing strains and allowing them to compete directly. Competitive fitness is measured by estimating the selection coefficient, s, directly as the rate at which an evolved population or strain eliminates the ancestor in head-to-head competitions. Regardless of whether the experiment was performed in batch culture, on plates, or in a chemostat, s is calculated in the same way:

$$\ln\left[\frac{x_E(t)}{x_A(t)}\right] = \ln\left[\frac{x_E(0)}{x_A(0)}\right] - st$$

where x_E is the density of the evolved strain (or population), and x_A is the density of the ancestor at time 0 or time t. Time is usually measured in generations, estimated as described earlier. Rearranging gives an expression for s:

$$s = \frac{\ln\left[\dfrac{x_E(t)}{x_A(t)}\right] - \ln\left[\dfrac{x_E(0)}{x_A(0)}\right]}{t}$$

This expression is in units of time, typically per generation or per day. The interpretation of s is thus straightforward: it is the rate at which one strain replaces the other. An $s > 0$ means that the competitor has higher fitness than the ancestor, and so adaptation has occurred. Consult Bell (2008) for a fuller description of the different procedures for calculating s. By convention, fitness is often reported as $w = 1 + s$, meaning that fitness has increased if $w > 1$.

It's important to realize that the estimation of fitness, either through growth parameter estimates or competitive fitness assays, assumes implicitly that fitness increases are additive. If this assumption is not met—for example, because non-transitive fitness interactions take place among strains, such as in a rock-paper-scissors scenario (see, e.g., Kerr et al. 2002) or when fitness is frequency dependent (as in the example from *Pseudomonas fluorescens* described later in the chapter)—the estimates of fitness become complicated and great care must be taken in interpreting the results. An example of how to estimate the frequency-dependent fitness is given later in the discussion of the model adaptive radiation in *Pseudomonas fluorescens*.

Replication

To distinguish treatment effects from experimental error, any well-designed experiment requires a certain degree of replication. This is no less true in a selection experiment, although it's not always obvious what to replicate. Is it better to have more independently evolving populations within a given treatment, or should the focus be on providing more accurate, and so more highly replicated, fitness estimates of one or a few independently evolved populations? Estimates of variance among independently evolving replicate populations include true experimental error due to, for example, uncontrollable variation in pipetting or plating efficiency. Such estimates also include error due to aspects of the evolutionary process itself, such as variation in the timing and identity of mutations substituted across different lines. The latter approach, obtaining replicate estimates of fitness from the same evolved strain or population, should include only true experimental error. The scope of inference is different in each case. In the former, one makes inferences about the efficacy of selection at generating adaptation in a given treatment. Does environmental variation prevent adaptation, for example, relative to a more homogeneous environment? Does sexual reproduction impede or promote adaptation relative to asexual reproduction? In the latter case, the inferences are made about the one evolved population being studied. This might be appropriate if, for example, you wished to study the genetic causes of adaptation in a given evolved line by estimating the fitness effects of mutations uncovered through whole genome sequencing.

MICROBIAL SELECTION IN ACTION

If the preceding discussion seems a bit abstract, then perhaps it is useful to turn to some concrete examples. Here are two that illustrate the main themes of this book, adaptation and diversification. This discussion is followed by a short section on bacteriophage experiments, because these studies also figure prominently throughout the book and are conducted in slightly different ways than the canonical experiments discussed next.

Long-Term Experimental Evolution in Escherichia coli

In 1988, Richard Lenski began what has become the longest-running intentional selection experiment on microbes. He started with a single colony of the common gut bacterium *Escherichia coli* B. Then, after isolating an isogenic mutant that had a mutation causing colonies to grow red instead of white on an appropriate indicator medium, he established six replicate populations of each strain in a minimal medium that contained a single carbon source, glucose. He and his lab have been transferring these twelve populations daily ever since. As of this writing, the populations have been evolving for approximately 60,000 generations.

Lenski's original objective was impressive for its simplicity: track adaptation and divergence among the replicate populations when founded from a common ancestor and propagated in a defined, homogeneous environment (Lenski et al. 1991). The strains used to found the experiment came from a single colony, so there was no standing genetic variation initially, and the samples are never dispersed among the replicates. This means that the only genetic variation available to the populations had to come from mutation. It's easy to predict that selection would generate adaptation through the substitution of beneficial mutations, but no one knew whether adaptation would be slow or fast, whether the lines eventually would cease evolving, or whether the replicate populations eventually would converge on the same adaptive solution or continue diverging indefinitely. This experiment was therefore poised to answer some of the simplest, but most fundamental, questions in evolutionary biology.

What Lenski and colleagues found was simultaneously mundane and astounding. First, why was it mundane? Because adaptation occurred, as did divergence. This was no surprise, because the starting clones were unlikely to be well adapted to the selection environment and mutations would be introduced at different times and in different places across the genome. The dynamics of adaptation and divergence over the first 10,000 generations are shown in **Figure 1.3** (Lenski and Travisano 1994).

Adaptation was rapid at first and then slowed down, revealing a decelerating pattern over time. Because all replicates were clones at the beginning of the experiment, divergence among lines necessarily occurred. This happened at a rate mirroring that of adaptation: fast at first, and then slowing down as time went on. Currently the lines still appear to be increasing in fitness, although at a much slower rate than they did in the first couple of thousand generations. They are still diverging, again at a much slower rate than they did initially. Thus, even in this most simple environment, consisting of a salt buffer and a single resource, these lines show no sign of converging on a single, universally superior adaptive solution. Evidently the mutations that occurred early on have had lasting effects on the genetic pathways these lineages have subsequently taken.

One of the more astounding results to have come from this work was the emergence of three lines out of the 12 with elevated mutation rates by the 10,000-generation mark (Sniegowski, Gerrish, and Lenski 1997). These so-called mutator lines increased in frequency because the mutations causing

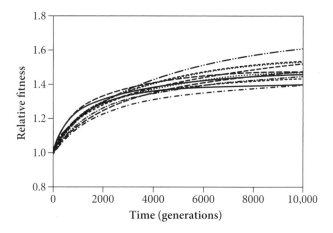

Figure 1.3 Trajectory of mean fitness relative to the ancestor (which by definition has a fitness of 1.0) for 12 independently evolving lines of *E. coli* over 10,000 generations. Note that fitness increases rapidly over the first 2000 generations and then more slowly after that. Average fitness reached by each line is different and shows no signal of convergence even after 10,000 generations. Reproduced with permission from Lenski and Travisano (1994) and Copyright (1994) National Academy of Sciences, U.S.A.

elevated mutation rates, which were typically loss-of-function mutations in DNA repair genes, produced beneficial mutations that caused adaptation (for a review, see Sniegowski et al. 2000). Another was the emergence, after about 35,000 generations, of a novel specialist genotype that metabolizes citrate and coexists with the glucose-specialist population. This result is quite incredible, because one of the diagnostic traits of this strain of *E. coli* is its inability to aerobically metabolize citrate. See Chapter 9 for a discussion of the genetic mechanisms responsible for citrate metabolism.

Many other fascinating results have emerged from this experiment, including data on maintaining diversity within populations, the dynamics of genomic evolution, and the evolution of trade-offs. These topics all find their way into the discussion that follows.

Diversification in Pseudomonas fluorescens

Pseudomonas fluorescens is a ubiquitous Gram-negative bacterium often found in association with the rhizosphere, or root system, of many plants. Its ecological role is unclear; but in plants at least, it is not strongly pathogenic. In the early 1990s, Paul Rainey was working with a strain of *P. fluorescens* that had been isolated from a sugar beet leaf from Wytham Wood, just outside of Oxford, U.K. Having left some cultures on the lab bench over the weekend, he noticed thick, sticky mats had formed across the top of the broth. This is not an unusual phenomenon. Many "wild" bacteria, when first cultured in the laboratory, often stick to the walls of a glass microcosm or clump together in the broth, making them hard to work with. As mentioned earlier, microbiologists often spend a few weeks or even months subculturing their samples to get well-behaved cultures—ones that grow nicely as single cells in the broth and do not clump or stick to the walls. Effectively, these researchers are selecting for domesticated strains, although most microbiologists will usually say they are "acclimating" their strains to lab culture. Whether this period of domestication/acclimation involves genetic or physiological changes is not normally investigated; it is simply standard practice in microbiology.

Rainey decided to investigate these poorly behaved cultures in more detail, rather than domesticate them. Working with Mike Travisano, he showed that the wild strain of *P. fluorescens,* known as SBW25, rapidly and repeatedly diversified over the course of a few days when grown in static (unshaken) microcosms containing a rich medium called King's B broth. Diversification occurs into a range of niche specialist types that occupy different regions of the microcosm and can be identified by eye based on their unique colony morphologies (Rainey and Travisano 1998). There are three predominant niche specialist classes, shown in **Figure 1.4a:** *smooths* (SM) that occupy the

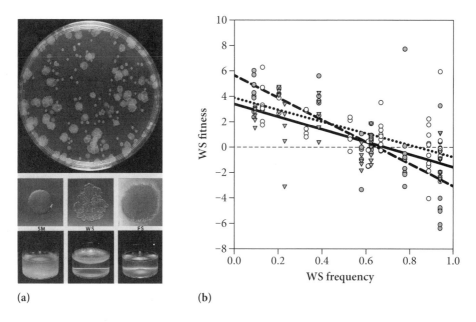

(a)   (b)

Figure 1.4 Adaptive diversification in *Pseudomonas fluorescens*. (a) Diversification in a static microcosm leads to the emergence and stable coexistence of three niche special-ist types with distinct colony morphologies. Pictured is a diluted culture of a diversi-fied population plated on agar (top). Middle and bottom panels depict the three main colony morphotypes and their niche preferences, respectively. SM: *smooth* morph that occupies the broth; WS: *wrinkly-spreader* morph that forms a biofilm and occupies the air–broth interface; FS: *fuzzy-spreader* morph that initially forms rafts and the air–broth interface and then falls and accumulates in the anoxic layer on the bottom. Photos generously provided by P. Rainey. (b) Diversity among colony morphs is maintained by negative frequency-dependent selection. The graph shows the relative fitness of WS as a function of WS frequency in competition with SM. The different lines represent differ-ent treatments. Redrawn from Meyer and Kassen (2007).

broth, *wrinkly-spreaders* (WS) that form a biofilm at the air–broth interface, and *fuzzy-spreaders* (FS) that were thought to be anoxia specialists but actually appear to be transient air–broth specialists that form loosely-adhering "rafts" of cells after a *wrinkly* biofilm collapses.

Diversity among these niche specialists is maintained by negative frequency-dependent selection: *smooths* can invade a population of *wrinklies*, for example, and the reverse is also true. This is shown in **Figure 1.4b**, which displays the relative fitness of a *wrinkly* type at different starting frequencies against the ancestral *smooth* on their own (dashed line), in the presence of a predator (solid line), and in the presence of a predator when the bacteria are prevented from growing by the presence of a low concentration of antibiotic (dotted line). In all cases the slope is negative, indicating that the fitness of WS is highest when it is rare and lowest when it is common. In this way, diversity is stably maintained and protected from loss.

Remarkably, diversification never happens if the microcosms are gently agitated on an orbital shaker rather than left undisturbed. Moreover, a previously diverse community will rapidly lose diversity if shaken. These results provide direct evidence that the spatial structure of the microcosm is supporting diversity. Rainey and colleagues suspect that competition for oxygen is the main driver of diversification in this system. Because oxygen enters the broth by diffusion, it quickly becomes limiting when the population size starts to increase after inoculation. The biofilm-forming *wrinkly-spreaders* presumably gain ready access to oxygen by remaining in the mat at the air–broth interface. Notably, *smooths* can often be found hitchhiking in the mat and, if they get too numerous, can lead to its collapse because they weigh it down further but contribute nothing to its structural integrity (Rainey and Rainey 2003; Kassen, Llewellyn, and Rainey 2004). Resources in the medium itself also seem to be important for diversification, because diversification doesn't happen in minimal media containing a salt buffer and a single carbon source. It's not clear which resources are important. King's B is somewhat like a liquefied cow, meaning that the resources it contains include a mess of amino acids, vitamins, and carbohydrates in unknown quantities and, sometimes, unknown identity. It would require something close to a lifetime of work to identify the key resource, or more likely resources, in the medium that facilitates diversification in this system.

By now, quite a lot is known about the genetics underlying the transition from broth-colonizing *smooth* to biofilm-forming *wrinkly-spreader*. **Figure 1.5** is a schematic of what we know about the routes to *wrinkly-spreader* formation.

Formation of the biofilm is governed by a two-component, signal-transduction-like mechanism. An as yet unidentified environmental signal is received by the first component, resulting in the production of a response regulator (cyclic di-GMP) that controls the expression of a structural operon (*wss*) and produces a cellulose-like polymer that forms the biofilm. A range of distinct mutations within one of three operons that receive the environmental signal, most often *wsp,* and that all negatively regulate genes encoding enzymes called diguanylate cyclases, leads to over-expression of cyclic di-GMP, and so to overproduction of the cellulosic polymer that forms the biofilm (Bantinaki et al. 2007; McDonald et al. 2009). Notably, these genetic changes are all loss-of-function mutations, at least at the level of the enzyme in which they are expressed.

The fact these are all loss-of-function mutations helps explain why the diversification is so rapid. Loss-of-function mutations are likely to be fairly

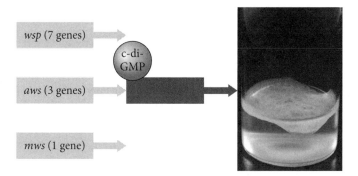

Figure 1.5 Cartoon of a genetic model for transition from broth colonizing *smooth* to biofilm-forming *wrinkly-spreader* in *P. fluorescens*. Genetic changes causing *wrinkly-spreader* (Photo by Andrew Spiers) typically occur as loss-of-function mutations in genes within one of three signal transduction operons (*wsp, aws, mws*) that negatively regulate the production of diguanalate cyclases, leading to the overproduction of c-di-GMP, a signaling molecule, that causes the *wss* operon to produce the cellulose-like polymer that forms the biofilm and provides access to the air–broth interface. (Adapted from Bantinaki et al. 2007 and McDonald et al. 2009.)

common, because they typically result from common errors in replication. To the best of our knowledge, the mutation rate in this strain, which is about 2×10^{-9} mutations per gene per generation, is not unusually high. But in combination with the large population size, multiplicity of genetic routes to the same phenotypic endpoint (biofilm formation), and strong selection for gaining access to oxygen at the air–broth interface, it is not hard to see how this diversification event could be so fast and repeatable.

Bacteriophage Experiments

It is worth including a few words about bacteriophages as models for experimental evolution, because the techniques involved are slightly different from those used for bacteria and other free-living microbes like yeast and algae. The main difference, of course, is that bacteriophages are viruses of bacteria and so require host cells for growth and reproduction. Normally, these host cells are treated as if they were components of the abiotic environment and not allowed to evolve, although a number of studies have begun to focus on phage-host coevolution (reviewed in Brockhurst et al. 2007b).

One of the most important differences to keep in mind when interpreting the results of phage experiments is that fitness is usually measured as the number of phage doublings per hour—rather than per generation, as is more common in bacterial and yeast experiments. The reason may in part be historical, because estimates of phage population growth rates are usually done over very short time scales, often 20 to 40 minutes or so; but it is also because fitness changes in phages are often massive, sometimes on the order of tenfold to a hundredfold differences over the course of a few days. A large selection coefficient (s) in population genetics is often taken to be anywhere from 0.01 to 0.1; in phage experiments, one phage study documented s to be 3.2–13.9 (Bull, Badgett, and Wichman 2000). These sorts of values are not exceptional in phage experiments, as we shall see. Such large differences in fitness must ultimately involve changes in generation time, making it difficult to provide a meaningful average generation time in phage experiments.

SUMMARY, AND A LOOK AHEAD

The principles behind experimental evolution are very simple. A population that is maintained in a perpetual state of growth will evolve by natural selection, operating either on preexisting genetic variation in fitness or on de novo variation arising through mutation. The effectiveness of natural selection at generating adaptation can be assessed using direct measures of fitness, preferably through competition between evolved and ancestral genotypes. If that is not possible for some reason, adaptation can be estimated by measuring key parameters of population growth when the genotypes are grown in isolation. Genetic diversity is estimated as the variance in fitness among isolates from the same population.

The experimental evolution approach also has a long history in biological research. Nevertheless, evolutionary biologists only recently have developed microbial experimental evolution as a stand-alone research program. A few of these experimental systems are described here. These experiments illustrate how experimental evolution in microbes is carried out in practice, and they provide a first look at the questions they are best suited to answering. The chapters that follow will emphasize these questions over and above the systems themselves. This is, after all, a book about the nature of biodiversity and how it evolves. Experimental evolution is a tool used to understand the process of diversification.

That said, it is important to be as clear as possible about what the tool can and cannot do. Experimental evolution is a powerful means of studying adaptation and diversification, because the experiments are conducted in a region of evolutionary parameter space where natural selection works most effectively. Population sizes are typically large, often on the order of 10^5 to 10^9, and experiments can last anywhere from a few tens to many thousands of generations. Genetic variation, in particular that generated through mutation, is not usually in short supply. Moreover, population sizes are so large that genetic drift, the random changes in gene frequency arising from small population sizes, can be effectively ignored. Reproduction is mostly asexual, and often the species are haploid. Taken together, these characteristics of microbial selection experiments predispose them to evolutionary responses driven by natural selection.

The value of microbial experimental evolution is therefore that it provides insight into how adaptation and diversification take place when natural selection is allowed to do its work. This is not to say that the complicating factors in evolution like drift, recombination, varying ploidy levels, and so on cannot be studied. As we shall see, many microbial experiments have investigated these and other mechanisms for how they contribute to adaptation and diversification. What experimental evolution provides is a means of testing the impact of these mechanisms on adaptation and diversification directly. This book is about the insights into adaptive diversification that have come from this approach.

CHAPTER 2

Adaptation to a
Single Environment

During World War II, the Nazis mounted regular bombings of major centers in Britain. Londoners sought shelter, at first against the will of the government, in the tunnels and stations of the Underground. Conditions were cramped and uncomfortable, and people were packed in like so many sardines in a tin. Adding to the discomfort, not to mention the fear, was a nightly assault of mosquitoes. Reports of mosquito infestations first came from the Liverpool Street Station but soon were arriving from all shelter points around the network (Shute 1951).

There was something puzzling, though, about the mosquitoes themselves. The offending species was found to be, in all morphological respects, identical to a common species found aboveground during the summer months: *Culex pipiens*. But *C. pipiens* was known to be a species that fed on birds, not mammals. Something was amiss.

Could it be that a new species of Tube-dwelling mosquito had evolved from an aboveground ancestor to invade this novel niche (Mattingly et al. 1951)? It certainly seemed possible, because belowground mammals like rats were in exceptionally high abundance, something that's still true today. Or was it that the belowground populations were actually distinct species that invaded a niche they were previously well adapted to? No matter. New species or not, the belowground populations received the uncertain, but painfully accurate, taxonomic status of *C. pipiens* var. *molestus*.

Further investigation revealed many other differences between aboveground and belowground populations of *C. pipiens*. Aboveground, *pipiens* bred in large swarms in open water, required a blood meal to produce eggs, and underwent a yearly diapause to make it through the cold British winters. The belowground populations managed to get around all these supposedly species-defining characteristics. They bred in small, confined pools of water underneath the station platforms, didn't require a blood meal to produce eggs, and the relatively constant, warm temperatures belowground meant they didn't need to go through diapause.

If belowground populations evolved from a local aboveground ancestor, adaptation would necessarily have been extremely fast. The first trains ran through the Underground in 1863, meaning that the *molestus* populations that so annoyed Londoners during the Blitz would have evolved from *pipiens* within less than 80 years. Even if the mosquitoes could breed three to five times a year, adaptation to life belowground would have taken somewhere between 500 and 1000 generations. This is the high-speed-train version of adaptive evolution, especially when we consider the many different traits that were involved. It is hard to square the magnitude and spectrum of character changes with what has become the standard model of adaptation involving gradual changes stemming from genes of small phenotypic effect (Fisher 1930).

The alternative hypothesis was that *molestus* were preadapted to conditions in the Underground because they had evolved elsewhere. One suggestion was that the evolution of a subterranean-like habit in this species happened alongside the agricultural revolution about 10,000 years ago, when aboveground mosquitoes first evolved the ability to breed in small, standing pools of water associated with farming in the warmer climates of southern Europe and North Africa. Populations of *pipiens* do exist in these warmer regions, and they don't require a blood meal to complete development. These variants would be partway toward making a go of life in the Underground. To complete the transition, the populations would need to switch their biting habits from birds to mammals. Members of the *pipiens* complex are known to do this occasionally (Mattingly et al. 1951), so the adaptive leap required likely wouldn't have been too great.

Thus the question of how *molestus* evolved really comes down to figuring out whether the rate of adaptation to life belowground occurred at a gallop or a trot (Byrne and Nichols 1999; Fonseca et al. 2004). It is notable that both scenarios would still be counted as fairly rapid evolution, at least on a geological timescale. It's also worth mentioning that the invasion by aboveground-like mosquitoes into underground transitways is not unique to London: *molestus*-like mosquitoes have been found in the subway systems of New York (Kent, Harrington, and Norris 2007), Chicago (Kothera et al. 2010), and Amsterdam (Reusken et al. 2010). All these species show evidence of being derived from aboveground populations, although the age of the relationships between above- and belowground populations in these cities remains unclear.

Whether adaptation is in general slow or fast has been an enduring question in evolutionary biology. The standard view since Darwin's time has been that it is exceedingly slow, proceeding through many, often imperceptibly small, changes. An argument for why adaptive evolution should be slow and gradual is easy to intuit. Most organisms are probably fairly complex beasts that have many traits important for fitness. If an improvement in one trait causes other traits to suffer, then large changes in many traits simultaneously are unlikely to improve adaptation; only small, relatively minor tweaks stand any chance of being favored by selection.

Yet rapid adaptation involving substantial changes in many traits does happen in organisms other than mosquitoes. The evolution of pesticide resistance, the invasion of freshwater by marine species, evolution on islands—in all these examples, adaptation has proceeded exceedingly fast and involved many changes in physiology and morphology to accommodate life in a new environment (Reznick and Ghalambor 2001). It's hard to justify dismissing these examples as exceptions to the rule that adaptive evolution is slow and plodding. It may be better to see them as one extreme of a continuum, and so refocus attention on understanding how a theory of adaptation can accommodate variation in rates.

That is the goal of this chapter. The central issue involves the magnitude of fitness increases and trait changes wrought by natural selection when it acts on beneficial mutations. Is most adaptation slow and gradual, being driven by very small changes in phenotype, or is it rapid and step-like, fueled by mutations of large effect? The answer to this question may often depend on whether the relevant mutation already exists in the population. When standing genetic variation is abundant the chances that the 'right' mutation is available are high, and so adaptation can proceed quickly. When selection acts on de novo genetic variation introduced by mutation, however, whether adaptation is slow or fast depends crucially on the fitness of the novel variant. The main obstacle to developing a theory of adaptation is therefore to understand how nature assigns a fitness value to a novel allele, and so this chapter emphasizes mutation, rather than standing genetic variation, as the source of genetic variation in fitness.

ADAPTATION: BASIC THEORY

Adaptation is a two-step process. The first step is the introduction of genetic variation. In an initially genetically uniform population closed to immigration, mutation is the sole source of genetic variation. If the population already contains substantial amounts of genetic variation, then, in addition to mutation and provided the population is capable of sexual reproduction, recombination will reshuffle preexisting variation and generate new genotypes.

The second step is natural selection. Genetic variation in the rate of replication among genotypes will cause the genotype with the highest rate of replication to come to dominate the population, all else being equal. Maladapted genotypes are maintained at a low frequency in the population by mutation and, if the population is open to dispersal, by migration as well.

These two processes, iterated many times, generate adaptation to the prevailing conditions of growth. The mean fitness of the population increases through time as successively better adapted genotypes are fixed. The process stops when all new genetic variation introduced by mutation results in a decrease in fitness. The population is then said to have achieved evolutionary equilibrium, at least in the context of the particular set of environmental conditions available.

This verbal theory is appealing for its simplicity but, at the end of the day, not very useful. All it says, really, is that adaptation happens when beneficial mutations are available. It tells us nothing about when mutation is likely to generate fitter variants or, if this does happen, how much more fit they will be. We need this kind of information to help make quantitative predictions about how much adaptation is likely to occur, and how fast. The gap in our understanding lies not with selection; the models of population and quantitative genetics have given us a sophisticated theory for how genes are substituted or lost from a population, and the consequences of these events for population-level measures of genetic variation. Rather, the gap comes from our inability to determine how nature assigns a fitness value to a mutation. This is the subject of the next section.

THE FITNESS EFFECTS OF BENEFICIAL MUTATIONS

Two models for understanding the origin of genetic variation through mutation, and how this affects fitness, have been put forward. The starting point for both is a large population currently fixed for a single genotype that is maladapted to the prevailing conditions of growth. Where the models differ is in the traits of primary concern: one models phenotypes, and the other models DNA sequences. The two models owe their original formulations to Fisher (Fisher 1930) and Gillespie (Gillespie 1984), and they have come to be known as Fisher's geometric model and the mutational landscape model, respectively.

In the simplest version of Fisher's model, we imagine that an individual is described by the particular combination of two traits, x and y, it possesses. Provided that the traits are genetically uncorrelated, the trait space in which evolution happens can be represented as a two-dimensional graph of x versus y (**Figure 2.1a**). Taking the optimal combination of traits (where fitness is highest) to be the origin, then the distance of any individual in trait space to the origin is a measure of the degree to which it is currently maladapted. To understand the effects of mutations on phenotype, imagine that we locate a

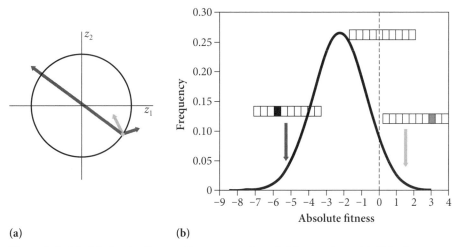

(a) **(b)**

Figure 2.1 Models for the fitness effects among beneficial mutations. (a) Fisher's geometric model. (b) Gillespie's mutational landscape model. Detailed descriptions in the text.

population in trait space as the gray dot in the lower right quadrant of Figure 2.1a. Suppose the average trait values of that population are x' and y'. Mutations are represented as vectors that change trait values of individuals. A beneficial mutation is one that moves an individual closer to the optimum. In the figure, this means that any mutation occurring within the circle is beneficial. Mutations that are deleterious move an individual away from the optimum; they represent combinations of trait values that lie outside the circle. The precise fitness value associated with a mutation depends on the function used to relate phenotype to fitness. Most often the function used is Gaussian.

Gillespie's model, on the other hand, considers mutations to DNA sequences rather than phenotypes. Imagine a gene of length L—1000 base pairs seems to be a reasonable value for many genes—and allow single base-pair mutations to occur anywhere along this gene. Given a particular starting wild-type sequence, there are $3L$ possible mutations that can be reached in a single mutational step (because a given site has only four possible bases, and one is already represented in the wild type). When a mutation occurs, we can determine its fitness by making a draw from some underlying distribution of fitness values (**Figure 2.1b**). The problem is, which distribution is the right one to use? Gillespie's insight was that beneficial mutations have, by definition, a higher fitness than the wild type. This means that when the wild-type population is not too low in fitness, beneficial mutations can be treated as draws from the right-hand tail of a distribution of fitness effects (the downward-pointing light gray arrow in Figure 2.1b; Gillespie 1984). The advantage of this approach is that the right-hand tail of many distributions look, superficially at least, quite similar. More importantly, though, the behavior of the tails of distributions can be characterized using extreme value theory, a branch of statistics that describes the probability of extreme and rare events such as floods, earthquakes, and market crashes. When applied to beneficial mutations in the way Gillespie described, extreme value theory allows us to make some compelling generalizations about the properties of beneficial mutations, and which of these mutations are available to selection.

Both models make predictions about what natural selection "sees" when it sorts among the genetic variance introduced by mutation. If we focus only on beneficial mutations, because they are the only ones that stand a chance of contributing to adaptation, then a simple qualitative prediction comes out of both models: most beneficial mutations are expected to have small effects on fitness, and few of them have larger effects. In other words, the distribution of fitness effects—the difference in fitness between a mutant and the wild type it is derived from—will be roughly L-shaped when the frequency of mutations of a given effect size are plotted against fitness. This prediction fits with our intuition about how nature works. If most organisms are fairly complex with lots of traits that can be changed by mutation, a random mutation is more likely to disrupt function than improve it. Only mutations of small effect on fitness, therefore, stand any chance of being beneficial.

An example distribution is shown in **Figure 2.2** (from Kassen and Bataillon 2006). For this work, mutants of *Pseudomonas fluorescens* that were resistant to an antibiotic called nalidixic acid were isolated. Nalidixic acid comes from a family of antibiotics, called quinolones, that interfere with the ability of DNA to replicate by binding to enzymes that help DNA wind and unwind appropriately. Resistance to quinolones comes from mutations in the binding pockets of these enzymes or from other mutations that act to pump the drug out of the cell before it interferes with DNA replication. The important point is that many different mutations within a gene, and among genes, can confer resistance. By selecting a large number of resistant strains that arise naturally during population expansion, the investigators were able to identify a wide range of genetically unique strains that usually differed by just one mutation from the ancestor (something later confirmed through direct sequencing; see Bataillon, Zhang, and Kassen 2011). When the fitness of these resistant mutants was assayed in an environment that lacked antibiotic, the expected decreasing relationship of many small-effect mutations and fewer large-effect ones was observed.

Table 2.1 summarizes the results of other, similar, experiments investigating the distribution of fitness effects among beneficial mutations. The result observed by Kassen and Bataillon is typical of most experiments with microbes that have examined the distribution of fitness effects among beneficial muta-

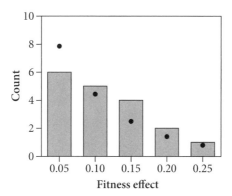

Figure 2.2 Distribution of fitness effects among beneficial mutations in *Pseudomonas fluorescens*. Bars represent observed frequency of mutations of a given fitness effect, and points represent those expected from an exponential model. (Adapted from Kassen and Bataillon 2006.)

tions. Thus the distribution of fitness effects among beneficial mutations is characterized by many mutations of small effect and few of large effect.

We can go a step further by making more quantitative predictions about the distribution of fitness effects. Beisel et al. (2007) have pointed out that the statistical theory used to describe extreme events like beneficial mutations actually describes three qualitatively different patterns for how the frequency of a mutation of a given effect falls off with the increasing magnitude of that effect. The most common pattern is one that decreases in an exponential fashion, with a long, asymptotic tail near zero. More technically, statisticians say that the limiting distribution of extreme values is in the Gumbel domain of attraction. Two other domains of attraction in this version of extreme value theory give different patterns. One is the Fréchet, which has a heavier tail than the exponential, meaning there are more extreme events on average. The other is the Weibull, which has a tail that is bounded on the right and so doesn't have an asymptote like the Fréchet or Gumbel.

If most distributions of fitness effects among beneficial mutations are decreasing, this discussion of the finer details of how they decrease may seem like splitting hairs. But there is one reason it might be important: if we can distinguish between distributions with and without right-hand bounds, then we can potentially say something about the nature of the underlying adaptive landscape. The most extensively studied version of Fisher's geometric model, for example, assumes a single phenotypic optimum, which naturally gives rise to a distribution of fitness effects that is bounded (assuming, of course, that the starting population we are studying is close enough to the optimum to detect such a bound). The distribution of fitness effects in Fisher's model is

Table 2.1 Tests of the Distribution of Fitness Effects (DFE) among Beneficial Mutations Available to Selection

Citation	Species	Environment	Mutations Isolated	Statistical Framework*	N_{ben}	DFE
Sanjuán, Moya, and Elena 2004	VSV virus	Baby hamster kidney cells	Site-directed mutagenesis	LS (exponential vs. gamma)	8	gamma
Kassen and Bataillon 2006	*Pseudomonas fluorescens*	LB	Threshold selection for nalidixic acid resistance	ML (exponential vs. gamma)	18	exponential
Rokyta et al. 2008	Microvirid DNA phage	*Pseudomonas syringae*	Threshold selection for host range mutants	GPD	16	Weibull (uniform)
	Escherichia coli		First fixed		9	Weibull (uniform)
MacLean and Buckling 2009	*Pseudomonas aeruginosa*	Low levels of rifampicin	Threshold selection for rifampicin resistance	GPD	9	Gumbel (exponential)
Schoustra et al. 2009	*Aspergillus nidulans*	"Complete" medium	Adaptive walks	NA	n/a	Weibull
Betancourt 2009	MS2 phage	*E. coli*	Adaptive walks	NA	n/a	Weibull
Bataillon, Zhang, and Kassen 2011	*Pseudomonas fluorescens*	55 carbon substrates	Threshold selection for nalidixic acid resistance	GPD	6–13	Weibull
Miller et al. 2011	Microvirid phage	*E. coli*	First fixed	GPD	16	Weibull (note exponential not rejected)

*LS: least squares; ML: maximum likelihood; GPD: generalized Pareto distribution.

thus necessarily in the Weibull domain (Martin and Lenormand 2008). It is worth noting that the mutational landscape model makes no explicit assumption about the connection between phenotype and genotype, so it cannot be used to make a priori predictions about the domain of attraction.

What do the data say about how often the distribution of fitness effects is Weibull, and so bounded on the right, versus otherwise? Table 2.1 is a summary of what is known to date. All studies have found the distribution of fitness effects among beneficial mutations to be either in the Gumbel or the Weibull domains; none are in the Fréchet domain. This result is somewhat reassuring because it tells us that even if there is no absolute limit to the size of a beneficial mutation, for most practical purposes massively large-effect mutations are so rare they will never be observed. More telling, though, is that six of eight distributions are Weibull, and so they are bounded on the right. The two studies where a Weibull was not found were actually those that did not even test for it, suggesting that a reanalysis of the data might tell a different story. This is indeed what happened in Kassen and Bataillon's study when they went back to reexamine their mutant collection and look at the distribution of fitness effects across a much wider range of environments (Bataillon, Zhang, and Kassen 2011): in all cases, they inferred distributions that were bounded on the right. Thus the evidence available points to a distribution of fitness effects among beneficial mutations before selection that is bounded, consistent with the idea that selection is happening in a phenotypic landscape with a single optimum.

However, the studies just described have an important limitation. Sample sizes are often so small that statistical power is weak, making it hard to distinguish among distributions. This is a largely unavoidable consequence of studying a phenomenon—beneficial mutation—that is very rare to begin with. Lack of power should not therefore be interpreted as a failing of the experimental methods but rather as a natural property of the data. An alternative approach is to generate all possible mutations across a gene (or genome) and investigate their fitness directly, thereby including both deleterious and beneficial mutations in the analysis. This technique has the attractive property of being a more comprehensive description of the mutations that selection might "see" in the initial steps of adaptation. High-throughput methods for doing this kind of experiment have been developed but have yet to be used for making quantitative inferences about the distribution of fitness effects among beneficial mutations (Hietpas, Jensen, and Bolon 2011). **Box 2.1** briefly describes the different methods used to study the fitness effects of beneficial mutations. These include traditional microbiological methods for isolating mutants (Kassen and Bataillon 2006; MacLean and Buckling 2009); small-scale, site-directed mutagenesis (Sanjuán, Moya, and Elena 2004); or inferring

---

BOX 2.1

Inferring the Distribution of Fitness Effects among Beneficial Mutations

Knowing the distribution of fitness effects among available mutations is crucial to developing a predictive theory of adaptive evolution, and the result has been a concerted effort by a number of groups to develop a sophisticated framework for inferring this distribution from data. This is not a trivial task: it involves finding the mutations themselves, estimating their fitness accurately, and then being able to infer an appropriate distribution. Four approaches have been used:

1. *Site-directed mutagenesis* Mutants are constructed directly by introducing known mutations into a wild-type genome. Often the number of beneficial mutations isolated with this approach is so low that no inference is possible.

2. *Threshold selection* Mutants are isolated using a strong selection procedure, such as expanded host range in bacteriophage or antibiotic resistance in bacteria. Under these conditions, the central assumption that justifies using extreme value theory is violated because the fitness of the wild type is zero and any mutation allowing growth under the selective conditions is beneficial. However, inferences can still be made either by treating the smallest effect mutation as a threshold (see discussion later in this box) or by estimating fitness of the mutants and the wild type under permissive conditions.

3. *First mutations fixed* The first mutations fixed in a large number of replicate selection lines are isolated. These mutations are a biased sample of available mutations, because they have escaped drift and possibly competition from other competing beneficial mutations and have been fixed by selection.

4. *Adaptive walks* This approach extends the procedure outlined for selection of first mutations fixed to multiple steps in an adaptive walk. By regressing the mean fitness effect at each step against the number of steps taken, inferences can be made about the domain of attraction in which the distribution of fitness effects among available mutants lies: no relationship implies the distribution is

the distribution from patterns of fixed mutations (Rokyta et al. 2008; Rokyta et al. 2005).

THE FIRST STEP IN ADAPTATION

A description, however crude, of the distribution of fitness effects among mutations available to selection is an important first step toward providing a more quantitative theory of adaptation. But it is by no means the whole story. Adaptation occurs only when a beneficial mutation increases in frequency in a population to a point where most or all individuals possess it. This is the process of substitution or fixation and, for it to happen, the beneficial mutation exponential and so is in the Gumbel domain, whereas a decreasing relationship implies it is in the Weibull domain. No stronger inference is possible, however. See Joyce et al. (2008) for more details.

As investigators we need to be able to make inferences about the shape of the distribution. This task is compounded by two factors. First, beneficial mutations are rare, and so sample sizes are inherently small. Increasing sampling effort, moreover, is unlikely to lead to significant increases in statistical power because typically the experiment just ends up increasing the chances of hitting the same mutations more than once—a form of genetic pseudo-replication. Second, most strategies for isolating beneficial mutations do not eliminate drift or selection, which introduces a bias toward mutations of larger effect. We thus are likely to miss the largest class of mutations, those of small effect.

To account for these problems, Beisel et al. (2007) have presented a statistical framework to infer the shape parameter of the generalized Pareto distribution. The generalized Pareto distribution describes the three domains of attraction in extreme value theory used to model beneficial mutations. Beisel and colleague's framework is designed for fitness data of the form collected in threshold selection and first mutations in fixed experiments. The version of extreme value theory used to generate the different domains of attraction involves a "peaks-over-threshold" approach, where beneficial mutations are defined as the peaks above the threshold fitness set by the wild type. Interestingly, inferences about the shape parameter in this approach are independent of the threshold level used. This is an attractive feature because it means that to infer the shape of the distribution of fitness effects, we do not need to know the wild-type fitness directly. Rather, we can take a more conservative approach by setting the threshold to be that of the smallest beneficial mutant observed and rescaling the data, sacrificing one degree of freedom in the process. Thus we can still make inferences about the shape of the distribution, even when we know there is a strong bias due to drift and selection against mutations of small effect. A maximum likelihood procedure for fitting data of this sort is available for the R statistical package from Paul Joyce's website at http://www.uidaho.edu/~joyce/lab page/computer-programs.html.

must first escape being lost from the population due to drift and then not be outcompeted by independently arising beneficial mutations in other individuals. The fact that a beneficial mutation arises is therefore no guarantee that a population will adapt. Here we take a look at the obstacles that beneficial mutations must overcome to contribute to adaptation as well as the properties of the beneficial mutations that do.

Drift

Drift happens because, by chance alone, some individuals contribute offspring to the next generation and others don't. It is a process that we usually associate with small populations, so it might seem odd that it plays any role at all in large populations like those typically studied in microbial experimental evolution. Yet drift does play a role because any mutation, including a beneficial one, is present in only a single individual when it first occurs. The effective population size of the mutation itself is therefore very small initially, and so drift will govern its dynamics while it remains rare. Recall that as a rule of thumb in population genetics, a mutation must have a selection coefficient greater than the reciprocal of the population size to avoid being lost due to drift. Beneficial mutations of small effect are therefore especially likely to be lost due to drift just after they arise; selection simply isn't strong enough to take over the dynamics of their substitution (Rozen, de Visser, and Gerrish 2002; Barrett, M'Gonigle, and Otto 2006).

Competition among Beneficial Mutations

More beneficial mutations are available to a large population than a small one. The reason is easy to see: if the genomic mutation rate to beneficial mutations is U_b per generation, then the number of beneficial mutations available is simply $U_b \times N$, where N is the population size. This combined quantity is known as the mutation supply rate. Sufficiently large populations are thus likely to produce many independently derived beneficial mutations that, provided they each escape drift, will compete with each other during the journey to fixation. In the absence of recombination, only a single lineage will fix, and these extra beneficial mutations are effectively lost. This effect, which is called clonal competition in asexual organisms and the Hill-Robertson effect in sexual ones, produces a further bias against small-effect mutations because only the mutations of largest effect will survive both drift and clonal competition.

The bias against small-effect mutations created by drift and clonal competition has two important consequences for adaptation. The first is that these events act to increase the average fitness effect of mutations that actually fix relative to what would be expected from the distribution of fitness effects among available mutations (Rozen, de Visser, and Gerrish 2002). This phenomenon is nicely illustrated by an experiment done by Schoustra et al. (2009). They found direct evidence for clonal competition in a selection experiment with the filamentous fungus *Aspergillus nidulans* (**Figure 2.3a**). One convenient feature of working with filamentous fungi is that beneficial mutations can actually be seen with the naked eye—they form pie-shaped sectors that outgrow the ancestral colony they are derived from. The colony pictured contains two beneficial mutations that have arisen independently (marked by the arrows) and would compete for fixation when the colony is washed off the plate and transferred. In Schoustra and colleagues work, these "contending" mutations—those that have escaped drift but have yet to fix— had mean fit-

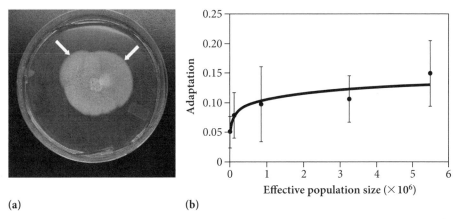

(a) **(b)**

Figure 2.3 Competition among beneficial mutations during adaptation. (a) A photograph of a five-day-old *Aspergillus nidulans* colony that has two beneficial mutations, indicated by the white arrows. These beneficial mutations form pie-shaped sectors that outgrow the parent colony. (Adapted from Schoustra et al. 2009.) Photograph by D. Gifford. (b) Rates of adaptation as a function of population size in asexual lineages of *Chlamydomonas reinhardtii*. Note the rate of adaptation increases rapidly with increasing population size in small populations and then levels off as population size increases, consistent with the effect of competition among beneficial mutations imposing a speed limit to adaptation. Adapted from Colegrave (2002). Reprinted by permission from Macmillan Publishers Ltd.: *Nature* 420: 664–66, © 2002.

ness effects that were on average 7 percent smaller than the mean of the first mutations fixed in the experiment. The bias against small-effect mutations caused by drift and clonal competition can thus make the first step of adaptation substantially larger than it would be without these phenomena.

The second consequence is that clonal competition creates a sort of speed limit on the rate of adaptation. Small populations founded from a single clone adapt slowly because beneficial mutations are rare events. A mutation like this doesn't often occur, and when it does, it is likely to be lost to drift because it probably has a very small fitness effect. So long as beneficial mutations remain limiting, the rates of adaptation will tend to increase with population size. Rates of adaptation reach a maximum, however, when population sizes are large enough that multiple beneficial mutations reach sufficient frequency that they compete with each other. Models suggest that clonal competition occurs when the mutation supply rate is about one beneficial mutation in every 10 generations (Gerrish and Lenski 1998).

An example of how clonal competition slows the rate of adaptation comes from work with *Chlamydomonas reinhardtii* (**Figure 2.3b**). In a strictly asexual population, the rate of adaptation first increases with population size but then reaches an asymptote at population sizes above about 1 million individuals. Such speed limits on the rate of adaptation have been observed in a wide range of model organisms, including phages (Miralles, Moya, and Elena 2000), bacteria (de Visser et al. 1999), and filamentous fungi (Schoustra et al. 2009). The speed limit can be broken by recombination, which brings together beneficial mutations that have arisen independently in different genetic backgrounds and so produce a genome with multiple mutations. In the *C. reinhardtii* experiment shown in Figure 2.3b, sex increased rates of adaptation by as much as 20 percent in the largest populations (Colegrave 2002). Such an advantage to recombination has been observed in other experiments as well (Goddard, Godfray, and Burt 2005; Cooper 2007).

Properties of Fixed Mutations

As mentioned in the previous section, one of the most important properties of fixed mutations is that, on average, they will have larger fitness effects than those that are initially available to substitution. The cause of this increase is, as discussed earlier, the bias against small-effect mutations caused by drift and competition among independently arisen beneficial mutations. With the one exception of the estimate provided by Schoustra et al. (2009), we don't have a good empirical handle on how much fitness increases due to the loss of these small-effect mutations. Models of adaptation can be used to calculate an expected increase in fitness, but doing so requires knowing the distribution of fitness effects among beneficial mutations (Rozen, de Visser, and Gerrish 2002). Until this distribution is better understood, the accuracy of our predictions about the extent of fitness increases during the initial stages of adaptation will be limited.

Other properties of fixed mutations are somewhat easier to evaluate, because natural selection is working in our favor. Fixed beneficial mutants that have escaped drift and clonal competition are necessarily abundant in the population, making them far easier to find. By isolating a large number of independently fixed beneficial mutations, we can therefore ask how the distribution of fitness effects among fixed mutations differs from that among available mutations. The expectation from theory is that the L-shaped distribution of fitness effects among available mutations becomes transformed as a result of the loss of small-effect mutations to drift and clonal competition into a distribution that has a humped shape for fixed mutations.

Table 2.2 summarizes what is known about the shape of the distribution of fitness effects among fixed mutations. Perhaps not surprisingly, these distributions are almost always hump-shaped, no matter which technique is used to isolate fixed mutations and the starting fitness of the ancestor. Even in those cases where the ancestor is strongly maladapted, and so the basic assumptions of both Fisher's model and the mutational landscape model are not expected to hold, the distribution of fitness effects is still hump-shaped. This result leads

Table 2.2 Summary of Studies Examining the Distribution of Fitness Effects (DFE) among Fixed Mutations

Citation	Species	Environment	Method	Replicate Selection Lines	DFE
Rozen, de Visser, and Gerrich 2002	E. coli	Glucose minimal medium	Marker ratio	30	Humped
Rokyta et al. 2005	Microvirid phage (ID11)	E. coli host	Fitness increase	20	Humped
Barrett, MacLean, and Bell 2006	P. fluorescens	Serine minimal medium	Marker ratio	68	Humped
Perfeito et al. 2007	E. coli	Luria-Bertrani rich medium	Marker ratio		
	$N \sim 1 \times 10^7$			10	Humped
	$N \sim 2 \times 10^4$			9	Humped
Schoustra et al. 2009	A. nidulans	Complete medium	ML		
	$N > 50,000$			60	Humped
	$N > 500$			60	Humped
MacLean, Perron, and Gardner 2010	P. aeruginosa	Rifampicin resistance	Resistance to high doses of antibiotic		
	A			96	Humped
	B			96	Humped
	C			96	Humped

ML: Maximum likelihood framework described in Schoustra et al. (2009).

to a very simple conclusion: the only mutations that are likely to fix are those of modest to large effect; small-effect mutations are simply wasted.

Two experiments have studied the effect of manipulating population size on the mean and shape of the distribution of fitness effects among fixed mutations. These studies are instructive for what they tell us about the effects of processes such as clonal competition on the mean effect size of mutations substituted. Perfeito et al. (2007) found that the mean fitness effect in large populations of *E. coli* was greater than in small populations, as would be expected if clonal competition were causing the loss of small-effect mutations in large populations. Schoustra et al. (2009) found the same result for the first step in adaptation in the filamentous fungus *A. nidulans,* but interestingly, the result did not hold for subsequent steps. Presumably, this happened because as the populations neared their adaptive peaks, the supply of beneficial mutations decreased sufficiently that clonal competition was effectively absent.

MULTIPLE STEPS IN ADAPTATION

In a few exceptional circumstances, a single mutation makes the difference between life and death; examples include resistance to an antibiotic or pesticide. However, most adaptation probably involves many genetic changes with more subtle effects on growth and reproduction. In populations that lack any standing genetic variation to begin with, all new genetic variation arises through mutation. The successive substitution of beneficial mutations by selection results in a process known as an adaptive walk, where each "step" is the fixation of one beneficial mutation. No new principles beyond those we considered in the preceding section are required to understand the individual steps. Rather, we need to focus on the number of steps required for adaptation to be complete and the dynamics of mean fitness that result.

The Length of an Adaptive Walk

How many mutations are typically substituted during an adaptive walk? Answering this question requires us to know when a walk has been completed. In principle this is easy. Each fixation event results in a new "wild type" that generates its own suite of mutations from which selection will choose. The process continues until no new beneficial mutations can be accessed by the current genotype. At this point, we say that the genotype has reached a fitness peak or optimum, and the adaptive walk is finished. In practice, of course, things are more complicated. We usually cannot detect mutations of extremely small effect, so it's difficult to tell if the population has reached a genuine adaptive peak or is just taking a very long time to get there. Moreover, it is always possible that the physical or genetic environment continually changes in a way that makes it effectively meaningless to talk about a population ever reaching a fitness peak, because the location of the peak keeps changing.

Nevertheless, within the parameters defined by these limitations—that almost all experimental methods will miss mutations of very small effect and that we focus on environmental conditions that remain reasonably stable in time—we can still make some predictions about how long an adaptive walk should be. What we predict depends on how much fitness is increased by a typical beneficial mutation. If most beneficial mutations have relatively small effects on fitness, at least relative to the distance to the optimum, then the lower the fitness of the starting genotype, the longer an adaptive walk will be, because it will take many small steps to reach an adaptive peak. This has, for much of the history of evolutionary genetics, been the standard view of how adaptation proceeds (Fisher 1930). A contrasting view, based on the two models of mutation introduced at the beginning of this chapter, is that beneficial mutations are highly variable in their fitness effects. Under this view, large-effect mutations will be preferentially fixed when they occur, thus allowing an evolving population to make up the extra distance otherwise covered by small-effect mutations. If mutational effects are highly variable, then adaptive walks will be fairly short.

Dynamics of Mean Fitness

These alternative views of mutational effects lead to two contrasting pictures of the dynamics of mean fitness during adaptation. The small-effect mutation model predicts a steady, gradual rate of fitness increase that is limited only by the availability of beneficial mutations. The variable effects model, however, generates a much faster rate of increase in fitness, followed by a leveling off as the fitness optimum is reached. This pattern arises in part from the reduced supply of mutations as a population approaches a fitness peak, just as in the small effects model, and in part because the fitness effect of the mutations substituted becomes smaller as the population nears an adaptive peak.

Data on Multiple Substitutions

What does the experimental evolution literature tell us about the length of adaptive walks and the dynamics of mean fitness? As an answer to this question, **Table 2.3** presents data gathered from the microbial selection experiment literature on the length of adaptive walks. Estimating the number of mutations fixed during adaptation has now become a relatively trivial task as whole genome sequencing becomes more accessible to many research labs. However, other methods have historically been used in microbial selection experiments, and these are discussed in **Box 2.2**. This data set includes only

Table 2.3 Number of Mutations Substituted (walk length, L) during an Adaptive Walk

Group	Species	Genome Size	Number of Genes	Citation	Selection Environment	Lineage Name	Generations	Δw	Harmonic Mean, N	L	Method	Notes
DNA viruses	phiX174	5386	11	Wichman et al. 1999	high temperature, *Salmonella*	Id	1000	2.4	10,000,000	14	Direct sequencing	Open-ended
	phiX174	5386	11	Holder and Bull 2001	high temperature, *E. coli*	n/a	150	8.3	19,980,019.98	3		
	G4	5577	11	Rokyta, Abdo, and Wichman 2009	*E. coli*	ID12a	180	3.8	1.10E + 05	2		
						ID12b	210	3.3	4.70E + 05	2		
						ID8a	240	9.8	8.30E + 04	5		
						ID8b	270	10.7	5.80E + 05	4		
						NC6a	210	4	3.50E + 05	4		
						NC6b	210	4.4	1.00E + 06	3		
	ID2	5486	11			ID2a	600	13.5	2.10E + 06	9		
	WA13	6068	11			NC28a	240	4.2	5.80E + 04	3		
						WA13a	120	9.3	6.30E + 05	2		
RNA viruses	phi6	13,385	13	Burch and Chao 1999	*Pseudomonas phaseolicola*	N10	100	0.26	19.99999998	2	Partial F tests	Compensatory
						N33	100	0.834	65.99999973	3		
						N100	100	0.815	199.9999975	2		
						N333	95	0.756	665.9999723	4		
						N1000	40	1.027	1999.99975	1		
						N2500	45	1	4999.998438	2		
						N10000	25	0.888	19999.975	1		

Table 2.3 (*continued*)

Group	Species	Genome Size	Number of Genes	Citation	Selection Environment	Lineage Name	Generations	Δw	Harmonic Mean, N	L	Method	Notes
Bacteria	*E. coli*	4,639,675	4497	Conrad et al. 2009	lactate	LactA	1100	0.71465	n/a	3	NGS	Open-ended
						LactC	1100	0.6487	n/a	3		
						LactD	1100	0.73571	n/a	7		
						LactE	1100	0.62571	n/a	7		
				Lee and Palsson 2010	propanediol	eBOP12	450	4.6	n/a	6		
	Pseudomonas fluorescens	7,147,633	6492	Bailey and Kassen 2012	mannose	M1	1000			2		
					glucose	G1				3		
						G2				3		
					xylose	X1				4		
						X2				4		
						X3				3		
Fungi	*Aspergillus nidulans*	30,070,000	10,560	Schoustra et al. 2006	minimal medium	WG621	3000	n/a	59,998.20	4	Partial F tests	Compensatory
				Schoustra et al. 2009	complete medium	WG615	800	0.48	99,995.00	2.39	Maximum likelihood	
								0.38	999.9995	2		
				Gifford, Schoustra, and Kassen 2011	complete medium	WG562A	960	0.888	99,995.00	2.05	Maximum likelihood	
						WG562B		1.081	99,995.00	2.15		
						WG562C		1.417	99,995.00	2.05		
						WG562D		2.868	99,995.00	2.4		
						WG562E		3.96	99,995.00	2.15		

BOX 2.2

Empirical Methods for Studying
Adaptive Walks

Identifying the mutations responsible for adaptation has been a major goal of evolutionary biology, and one of the advantages touted for microbial experiments has always been the ability to dissect phenotypic changes down to their genetic causes. That said, the genetic causes of adaptation have remained largely a mystery until the introduction of low-cost whole genome sequencing in the past few years. A related question that has proven modestly more accessible to empiricists is how many genes are involved in adaptation, and what their associated effects on fitness are. A range of methods to answer both questions have been proposed over the years, some of which were developed initially for sexually reproducing organisms and others that are unique to asexually evolving microbes. Here is a brief overview of these techniques.

Ratio of Variances in Sexual Organisms This method was initially developed for use in quantitative genetic analyses. It has since expanded to incorporate molecular markers as well, where it is called quantitative trait locus (QTL) analysis. The basic idea is to infer the number of genetic factors contributing to a phenotype by comparing the ratio of variances in a marker or phenotype between parents and their sexually reproduced offspring. The larger the offspring relative to the parental variance, the more genetic factors are involved in adaptation. The most familiar technique is the Castle-Wright method (Castle 1921; Otto and Jones 2000; Zeyl 2005), which was designed explicitly for selection experiments. These analyses tend to miss mutations of small effect. Details can be found in Lynch and Walsh's book, *Genetics and Analysis of Quantitative Traits* (Lynch and Walsh 1997).

Whole Genome Sequencing This is the most direct approach; but its utility, although growing as costs come down, remains largely limited to microbial

those studies where it is reasonably clear that the bulk of adaptation to prevailing conditions has been accomplished. In practice, this means that the study needed to provide statistical evidence that the evolving lines had reached a fitness plateau. This criterion eliminated many recent whole genome resequencing studies where a plateau had clearly not been reached. It also eliminated Lenski's long-term lines of *E. coli,* which have not reached a fitness plateau even after 50,000 generations of selection (Barrick et al. 2009). Also dropped from the analysis were other experiments where there is evidence of perpetual adaptation due to non-transitive fitness increases through time (e.g., Paquin and Adams 1983). Finally, it is always possible in the studies included here that further small-effect mutations could be substituted given sufficient time.

species with relatively small genomes. The idea is to compare the DNA sequence of evolved genomes against an unevolved ancestral genome to identify mutational changes. Current technologies are extremely good at identifying single nucleotide polymorphisms (SNPs), but larger structural changes such as rearrangements, duplications, and deletions can be missed. This method, and the insights it has provided for studies of adaptation, has been reviewed by Dettman et al. (2012) and is discussed at length in Chapter 5.

Inferences from the Dynamics of Fitness When the starting population is initially isogenic and beneficial mutations are sufficiently rare, a substitution event can sometimes be inferred directly as a steplike increase in fitness through time (Burch and Chao 1999; Lenski et al. 1991; Schoustra et al. 2006). A more sophisticated approach that does not rely on the detection of obvious steplike increases in fitness uses a maximum likelihood framework to fit models of population growth given different numbers and effect sizes of mutations (Gifford et al. 2011; Schoustra et al. 2009). Neither method is capable of distinguishing multiple mutations in the same genome fixing together from a single beneficial mutation. And, like variance ratio methods, they may miss mutations of very small effect.

Marker Frequency Dynamics This method is a variant of the marker ratio technique mentioned in Box 2.1. The difference is that now one marker is very rare relative to the other, rather than both being at equal frequency initially (Atwood, Schneider, and Ryan 1951; Paquin and Adams 1983). Under these conditions, the rare marker will increase in frequency through mutation alone and a beneficial mutation, when it occurs, is far more likely to happen in the more abundant marker type. The result is a sharp decrease in the frequency of the rare marker upon substitution of the beneficial mutant. Over time, successive substitutions produce a sawtooth pattern of the rise and fall of the rare marker, and the number of declines in the rare marker can be used as an indication of the number of selective sweeps that have occurred.

It is therefore appropriate to treat these measures of walk length as minimum estimates.

Let's start with the dynamics of mean fitness, because it is the easiest to assess. All the studies presented here share a similar dynamic pattern: fitness increases rapidly at first and then levels off. A typical result, from research work with evolving populations of *Aspergillus nidulans,* is shown in **Figure 2.4a**. This pattern is, of course, a necessary consequence of the requirement that evolutionary equilibrium be reached for the study to be included in this table. However, the really interesting question is why this decline happens. Is it because all mutations are of small effect and the supply of them runs out? Or is it because, in addition, large-effect mutations tend to be substituted first, followed by small-effect ones? Schoustra et al. (2009) provide direct evidence from their work with *A. nidulans* that beneficial mutations do indeed become rarer as adaptation proceeds. Moreover, at least three studies have shown that the first mutations fixed tend to be of large effect and the later ones substituted are of smaller effect (Betancourt 2009; Gifford, Schoustra, and Kassen 2011; Schoustra et al. 2009). Thus the decreasing rate of adaptation is due both to changes in the fitness effects of mutations fixed and their becoming increasingly hard to come by.

It is worth asking more about the causes of the small-effect mutations that get substituted later in adaptation. Do they have genuinely small fitness effects, or do they just appear to because they occur in the genetic background of a genotype with high fitness? The available evidence to date favors the latter view, a phenomenon known as diminishing-returns epistasis. Chou et al. (2011), for example, showed that in *Methylobacterium extorquens* evolved in batch cultures containing methanol as a sole carbon source, the fitness increase associated with four beneficial mutations that had substituted over the course of 600 generations declined as the fitness of the background genotype increased. Similar results have been observed in *E. coli* (Khan et al. 2011) and *Pseudomonas aeruginosa* (MacLean, Perron, and Gardner 2010).

What about the number of substitutions involved in adaptation, that is, the number of steps taken during an adaptive walk? **Figure 2.4b** shows a his-

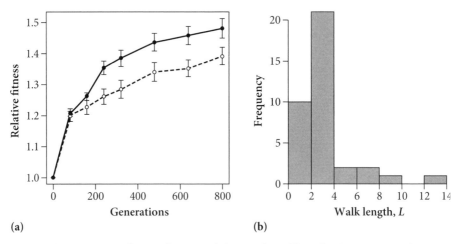

(a)   **(b)**

Figure 2.4 Dynamics of mean fitness and the number of beneficial mutations substituted (walk length, L) in microbial evolution experiments. (a) The rate of fitness increase declines with time in evolving populations of *A. nidulans*. Shown is the mean (± 1 standard error) fitness of approximately 60 populations propagated under large (solid circles and line) and small (open circles and dashed line) effective population sizes. (Data originally published in Schoustra et al. 2009.) (b) Histogram of L for all studies included in Table 2.3.

togram of length of adaptive walks from the studies in Table 2.3. The most striking result is that adaptive walks seem to be fairly short. They are typically between one and four steps, and the median is three steps. There are a few examples of longer walks, in particular one in phage that involved 14 substitutions. Long adaptive walks tend to be associated with massive increases in mean fitness, because the initial genotypes were actually declining in population size (Holder and Bull 2001; Wichman et al. 1999) or grew very slowly (Lee and Palsson 2010). If these populations are far from a fitness optimum, then presumably the supply rate of beneficial mutations is quite large, and it is almost certainly larger than the number of mutations ultimately fixed. Viewed in this light, even these longer walks might be construed as being fairly short.

Open-Ended versus Compensatory Evolution

One possible criticism of some of the adaptive walk studies that have been done—including my own—is that they are actually studies of compensatory evolution rather than truly open-ended adaptation. The distinction between the two comes down to the cause of maladaptation. Some experiments study adaptive walks by manipulating the external environment in a way that the founding population is known to be maladapted. This is often interpreted as open-ended evolution. Others generate maladapted founders by fixing a deleterious mutation and then allowing adaptation to proceed to "compensate" for this loss of fitness. This is compensatory evolution.

Do adaptive walks differ in any fundamental way during open-ended versus compensatory evolution? To examine this question in more detail, let's take a closer look at the experiments reported in Table 2.3 and ask whether there is a relationship between walk length and the total gain in fitness (s_{eq}, see Table 2.3 for details; **Figure 2.5**) between those experiments that have generated maladaptation through genetic or environmental means. It turns out there is. Compensatory evolution experiments—where the bulk of adaptation involves fitness recovery thanks to the prior fixation of a deleterious mutation that knocks a genotype off a fitness peak—show no relationship between walk length and the amount of fitness gained (slope $= -0.30 \pm .44$ standard error; $F_{1,13} = 0.47$, $P = .51$). There is a significantly positive relationship between walk length and fitness gain, however, for the experiments considered to be more open-ended (slope $= 1.70 \pm .58$ standard error; $F_{1,18} = 8.56$, $P = .01$).

Why this difference? The most likely reason stems from differences in the nature and number of beneficial mutations available. Compensatory mutations are, by definition, beneficial only in the presence of the deleterious mutation, and so they are unavailable during more "open-ended" evolution. If only a handful of compensatory mutations are capable of restoring fitness, adaptive

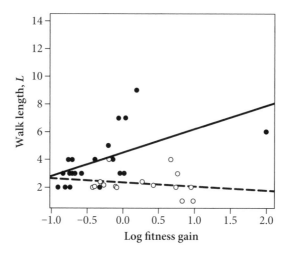

Figure 2.5 The length of adaptive walks, *L,* in open-ended (filled circles, solid line) versus compensatory (open circles, dashed line) evolution experiments as a function of the amount of fitness gained on a log scale. Data are from Table 2.3.

walks under compensatory evolution will be necessarily shorter than under open-ended evolution. This happens because the range of mutational pathways available for adaptation is much more restricted than in open-ended evolution. Evidence to support this interpretation comes from examining the number of mutations substituted in the two kinds of experiment. Though both are still quite short, adaptive walks for open-ended evolution are modestly longer when compared to compensatory evolution (median L_{open} = 3 steps vs. median $L_{compensatory}$ = 2.05 steps). Thus, fewer genetic routes to adaptation appear to be available during compensatory as compared to open-ended evolution.

Ruggedness of the Adaptive Landscape

An adaptive landscape relates how genotype, or sometimes phenotype, maps to fitness. The basic idea is that if there is a single fitness peak in the landscape, there will be a single, optimal genotype that sits on top of it, one that has the highest possible fitness. Selection acts to push populations up the slope by substituting beneficial mutations. It's also possible that the adaptive landscape could be very rugged, with many peaks and valleys. The array of peaks presumably represents the range of genetic routes that adaptation can take. Adaptive walks are expected to be shorter on rugged landscapes than on smooth ones, simply because a population randomly placed on a rugged landscape is more likely to be closer to a local fitness peak.

This prediction awaits a strong experimental test. It is, admittedly, a difficult experiment to run because one needs to define a priori which environments are rugged and which ones are smooth. There is a growing body of evidence that many selection environments are at least modestly rugged. Weinreich et al. (2006), for example, have shown that of the 120 possible mutational routes among five mutations known to be involved in resistance by *E. coli* to the antibiotic beta-lactamase, 18 were evolutionarily accessible in the sense that resistance (and so presumably fitness) increased at each step. In this system, then, there is no single optimal genotype for resistance. Moreover, studies that have tracked the variance in fitness among evolving lines during an adaptive walk show that this variance remains high even after fitness plateaus (Lenski and Travisano 1994; Gifford, Schoustra, and Kassen 2011; Schoustra et al. 2009). Presumably, the underlying genetic cause is negative sign epistasis—certain genetic combinations are simply not compatible with each other. Chapter 5 discusses a particularly compelling example of sign epistasis (Kvitek and Sherlock 2011).

ADAPTIVE WALKS UNDER DIFFERENT GENETIC SYSTEMS

So far we have been considering the simplest of all genetic systems—asexual haploids. However, many of the species people care most about are sexual and often diploid. How do the mode of reproduction and ploidy level affect adaptation involving multiple substitutions? Let's consider these issues here.

Recombination

Recombination should not affect the total length of an adaptive walk, but it can change the rate at which a fitness plateau is achieved. If mutation supply rates are sufficiently high that beneficial mutations arise independently in different lineages, then recombination among those lineages can bring mutations together and the result will be faster rates of adaptation.

There have been no direct tests of the effects of recombination on the number of mutations fixed during adaptive walks. However, a handful of studies have tested the prediction that recombination will speed the rate of adaptation. These were mentioned previously in the context of understanding how recombination could break the speed limit on adaptation imposed by competition among beneficial mutations in asexual populations (Colegrave 2002; Cooper 2007; Goddard, Godfray, and Burt 2005). Goddard, Godfray, and Burt (2005) showed, for example, that sexually reproducing populations of yeast adapt faster to high temperatures than asexual populations do, but this advantage disappears when both sexual and asexual lineages are grown under benign conditions they are well adapted to. This result is consistent with the idea that an advantage of recombination is its ability to speed up adaptation by bringing together distinct beneficial mutations into a single genome.

Of course, just as recombination can bring together beneficial mutations arising in different genetic backgrounds into the same genome, it can also separate them. This is one of the costs of sexual reproduction; the other costs stem from the time and energy associated with gamete production, investment in finding mates, and the processes of cell fusion and genome replication. Consistent with this idea that recombination can break up advantageous combinations of genes, a single round of sexual reproduction, for example, typically decreases the mean fitness of asexually adapting lines of the unicellular algae *Chlamydomonas reinhardtii* (Colegrave, Kaltz, and Bell 2002; Kaltz and Bell 2002).

An ideal strategy might be to invest in sexual reproduction only when fitness is low, because these are the conditions most likely to allow recombination to speed adaptation, but not when fitness is high, because recombination will then break up existing beneficial combination of genes. Schoustra et al. (2010) directly demonstrated that *Aspergillus nidulans,* a facultatively sexual filamentous fungus, uses just this strategy, a phenomenon termed fitness-associated sex. The evolution of sexual reproduction from an initially asexual ancestor may thus be easier to understand if it happened by way of fitness-associated sex as in the *A. nidulans* example (Hadany and Otto 2007).

Ploidy

The life cycles of most eukaryotic species are characterized by an alternation between haploid phases, with just one set of chromosomes, and diploid phases, with two sets. Exceptions exist, of course. Many plant species are polyploid, and the relative time spent at different ploidy levels during the life cycle varies quite a lot among species (Otto 2007). Nevertheless, ploidy level is often taken to be a good diagnostic trait of many species, thus suggesting it is a trait that has been selected and raising the question of how it might affect the rate and extent of adaptation.

It's easy to see that having more genetic material can increase the supply of beneficial mutations, all else being equal. If the supply rate of beneficial mutations is NU_b in haploids, for example, it will be $2NU_b$ for diploids, $4NU_b$ for tetraploids, and so on. In principle, this means that populations with higher ploidy levels could explore more of the mutational landscape for a given population size and so may be able to find that rare, large-effect beneficial mutation that a comparable population with lower ploidy levels cannot.

The result might be expected to be shorter adaptive walks in higher ploidy populations. In practice, though, the fate of a beneficial mutation depends, as we have seen, on the probability of fixation, which for a diploid is determined by what might be called the "effective" selection coefficient (hs), where s is the selection coefficient in a haploid (or homozygous) state discounted by the coefficient of dominance, h. The result is that the effect of ploidy on adaptive walks depends on both population size and dominance: only when beneficial mutations are rare ($NU_b \ll 1$) and dominant (formally, $h > 1/2$ for a diploid population) will diploids have a faster rate of adaptation than haploids (Orr and Otto 1994). Otherwise, haploids should have the advantage under most conditions.

In a particularly elegant and direct test of this prediction, Zeyl, Vanderford, and Carter (2003) contrasted the rate of adaptation in large and small populations of haploid and diploid yeast growing in a minimal medium with dextrose as a sole carbon source. Haploids adapted faster than diploids when population sizes were large ($N_e \sim 1.3 \times 10^7$) and at the same rate as diploids when population sizes were small ($N_e \sim 1.4 \times 10^4$). Interestingly, those mutations that fixed in diploids tended to be more dominant than those that fixed in haploids, consistent with theory. An advantage to haploidy over diploidy has been seen in other selection experiments with yeast across a range of growth conditions (Anderson, Sirjusingh, and Ricker 2004; Gerstein et al. 2011; Gerstein and Otto 2009).

Our understanding of how ploidy affects adaptive evolution is far from complete, however. For example, theory has not taken account of the variability in fitness effects among beneficial mutations that comes out of the mutational and phenotypic models. It's conceivable that the range of conditions favoring diploidy might be broadened somewhat if the distribution of fitness effects among beneficial mutations were such that it included extremely large-effect mutations or allowed the fixation of multiple mutations that in combination confer a larger fitness than any single mutation (see, e.g., Thompson, Desai, and Murray 2006). Moreover, most models of ploidy evolution assume that mutational effects are the same at different ploidy levels, although recent data from experiments with yeast suggest that beneficial fitness effects are on average larger in haploids than diploids (Gerstein 2013). We also do not know much about the distribution of dominance coefficients before selection. If h is on average less than $1/2$ to begin with, then the evolution of diploidy becomes more difficult to understand. Finally, we know very little about how the life cycle evolves and what maintains variation in the fraction of time spent as a haploid or diploid (Bell 1997a). This seems like a line of enquiry that is ripe for experimental investigation (Coelho et al. 2007).

It's important to understand that ploidy is a characteristic that itself can, and often does, evolve. Initially haploid and tetraploid yeast ended up being diploid following close to 2000 generations of selection in standard yeast medium, both under permissive conditions and stressful ones (Gerstein et al. 2006). The authors attributed this to a history of selection as a diploid in yeast. Complicating matters even more, diploid populations of the filamentous fungus *Aspergillus nidulans* that had reverted to haploidy during the course of a 3000-generation experiment had higher rates of adaptation than haploids themselves (Schoustra et al. 2007). Intriguingly, this result was due to the accumulation of recessive beneficial mutations that then underwent mitotic recombination during haploidization to generate high-fitness genotypes. The implication is that the so-called parasexual cycle in fungi, which involves chromosomal reassortment without meiosis, may be more important in driving adaptation than fungal biologists had previously thought. This result deserves more exploration, at least in terms of understanding the evolution of ploidy and sexuality in fungi.

ADAPTATION FROM STANDING VARIATION

The discussion so far has focused on one of the simplest imaginable models of adaptation, that all new genetic variants are introduced solely by mutation. This sort of scenario is obviously appropriate for thinking about adaptation in laboratory populations of microbes, and it may be useful for interpreting adaptation in natural microbial populations as well, like in *P. aeruginosa* infections of the cystic fibrosis lung (see, e.g., Folkesson et al. 2012; Smith et al. 2006; Wong, Rodrigue, and Kassen 2012). But it is almost certainly not appropriate for thinking about adaptation in many higher organisms where standing genetic variation is much higher and population sizes much smaller than in microbes.

How does adaptation proceed when standing genetic variance in fitness is large? The most obvious difference with a mutation-driven model is that the response to selection will be faster when selection uses standing variation than when it relies on mutation, simply because more genetic variance is available. This will be true especially when selection happens over short time periods in small populations. Beneficial mutations are simply unlikely to arise in these situations, and so they will contribute little if anything to the response to selection.

When population sizes are larger and selection happens over longer time periods, however, beneficial mutations may not be so rare. The question then is whether adaptation proceeds more often through changes in the frequencies of preexisting alleles or through novel variation introduced by mutation. Intuitively, it seems that selection from standing variation would be more likely, because preexisting genetic variation, even if it is being held at low frequencies by mutation–selection balance, is substantially closer to escaping drift loss than a beneficial mutation that is initially present in only a single copy.

In a model where a previously deleterious allele, segregating at mutation–selection balance, now finds itself beneficial after an environmental change, adaptation is indeed more likely to proceed from standing genetic variance rather than mutation (Hermisson and Pennings 2005). But it turns out that, in this model at least, in the time it takes for this allele to reach fixation, a comparable allele could arise from mutation and eventually be substituted. Thus it is not immediately obvious which source of variation is most important for adaptation.

We know from the success of quantitative genetics in animal and plant breeding that standing variation is extremely important in the response to artificial selection. It's clearly not pushing matters too far to say that standing variation will be important in governing adaptation in natural populations as well (Schluter et al. 2010). It appears that only one study involving laboratory selection has studied the dynamics of standing variation in detail (Teotónio et al. 2009), showing that phenotypic divergence in key life-history traits such as age of reproduction and starvation resistance is associated with changes in allele frequencies and not the introduction of new mutations, at least on time scales of a few hundred *Drosophila* generations. Clearly more work is needed, on both the theoretical and experimental side, before we will understand the relative roles of standing variation versus mutation in governing adaptation in natural populations.

EVOLUTIONARY RESCUE AND
THE DEMOGRAPHY OF ADAPTATION

The models of adaptation discussed here ignore the effect of environmental change on population size. Rather, they assume that adaptation proceeds from a founding population that is maladapted to current conditions, but not so maladapted that it cannot replace itself and goes extinct. When the environment changes so drastically that absolute fitness falls below one, however—as can happen under any number of scenarios such as the administration of antibiotics or pesticides, certain kinds of climate climate change, or the invasion by a pathogen of a novel host—population sizes will decline to extinction unless the population adapts in the meantime. Population survival therefore depends on the race between adaptation and demography (Maynard Smith 1989). "Evolutionary rescue" is said to occur if adaptation wins the race because a beneficial mutation restores absolute fitness to one or above before extinction wipes out the population (Orr and Unckless 2008, Bell 2008).

Bell and Gonzalez (2009) provide a particularly elegant experimental demonstration of evolutionary rescue, showing that when yeast, *Saccharomyces cerevisiae,* is challenged with a normally lethal level of salt stress, population sizes decline rapidly in the first day, remain at low level for another day, and then begin to recover over the next three days, or approximately 25 generations. Adaptation to a severe stress can thus occur rapidly and is accompanied by characteristic U-shaped changes in population size over time.

CONCLUSIONS

The invasion of subterranean transitways by previously free-living, aboveground mosquitoes is a particularly striking example of rapid adaptation to a novel niche. Although the evidence is equivocal on whether invasion more often happens de novo from local aboveground populations or through dispersal of partially preadapted individuals from elsewhere, in both situations adaptation must have been very fast. The model of adaptive evolution that was most widely accepted during the twentieth century sees adaptation as happening gradually and being driven ultimately by many genes of small effect. This model thus cannot, on its own, explain this or any other instance of rapid adaptation. It is interesting, though, that a simple modification to this standard model that incorporates a reasonable model for variation in the fitness effects among beneficial mutations goes a long way to explaining how and why adaptation can occur so quickly. Provided the starting genotype is maladapted to prevailing conditions, which the initial dispersers into the underground must have been, adaptation can proceed fairly quickly by substituting rather few genes of large effect, at least in the early stages of adaptation.

The results of the experiments reviewed in this chapter are entirely consistent with this view of adaptation. They can be summarized as follows:

1. The distribution of fitness effects among beneficial mutations before selection is roughly L-shaped and has many small-effect and few large-effect mutations.

2. This distribution gets transformed through the action of genetic drift and clonal competition into a modal distribution of fitness effects fixed by selection that has a substantially higher mean.

3. The bias against fixing genes of small effect means that the bulk of adaptation can be accomplished quickly; the number of beneficial mutations substituted is typically just a few, and the first mutations substituted tend to have larger fitness effects than those that follow.

4. Thanks in part to this pattern of large-effect mutations first, followed by small-effect mutations, but also due to the increasing rarity of beneficial mutations in high-fitness populations, the rate of increase in fitness over time is initially rapid and then slows down toward a plateau when external conditions are relatively stable through time.

5. Adding further biological realism into this model, such as accounting for recombination or ploidy, is not expected to change this conclusion.

6. Adaptation in small populations inevitably proceeds from standing variation, but in larger populations, both standing variation and mutation can potentially contribute to adaptation. A direct demonstration that both factors contribute to adaptation has yet to be provided.

This last point will be particularly important for most readers. Our enthusiasm in applying these models of adaptation due to mutation to "real" examples of adaptation in nature, including that of *molestus* adapting to belowground environments, must be tempered with a healthy dose of skepticism. The theory, and the experimental models used to test predictions so far, all involve adaptation from de novo mutations, not standing variation. Depending on the amount of genetic variation contained in natural populations that selection can sort from initially, we might expect these conclusions to change somewhat. At the very least, we would expect the response to selection to be faster when proceeding from standing variation as opposed to mutation, especially when population sizes are smaller than they typically are in microbial experiments ($N < \sim 10^5$). Nevertheless, what theoretical work has been done leads us to expect that both standing variation and new mutations could contribute to adaptation when both are available. It remains to be seen which of these factors is more important.

The conclusion that most of adaptation occurs through relatively few genes of large effect seems at odds with what we know about many ecologically important traits in metazoan systems (see, e.g., Rockman 2012). The history of quantitative genetics, for example, is rife with instances of successful artificial selection, often in opposing directions, that proceeds gradually for some time in a manner consistent with the traits being underlain by many genes of small effect. There are two possible explanations for this apparent paradox, neither of them mutually exclusive.

1. Because large-effect mutations fix quickly and small-effect ones take longer, the bulk of standing variation may in fact be due to genes of small effect, and some of those may even be modifiers of the large-effect mutations that previously fixed. This proposal could explain nonzero heritabilities for many traits and a continuous response to artificial selection, especially if many or most traits are under stabilizing selection. If these small-effect mutations are small enough that they rarely fix, then most adaptation will be due to changes in allele frequencies, as seen in the *Drosophila* example mentioned earlier, rather than fixation.

2. Most environments are not static, as the models and experiments analyzed here assume, but instead can be described through a model where the optimum is slowly but persistently moving. This proposal, as we will see in Chapter 3, tends to favor the fixation of smaller effect mutations.

Reality may, of course, be somewhere between these two extremes.

Divergent Selection

I grew up in British Columbia, on the Pacific coast of Canada. The coast is an idyllic place. It's wetter than most Canadians prefer, but compared to the rest of the country, BC winters are not as cold and summers are not as hot. Life is run not by the weather so much as by the sea. And, for BCers, salmon rule the sea.

The Nuxalk people of the Bella Coola valley, on the central coast, named the seasons and villages after salmon: Snut'lh, Place of Dog Salmon; Nutsats'm,

Figure 3.1 Sockeye salmon spawning in the Adams River, British Columbia. Photo by Matt Casselman.

place of the Biggest Spring Salmon; Sinumwak, the Moon for Making Salmon Weirs; Si7ist'lilhh, Dog Salmon Moon.[1] Dotting the coast are old canneries that made tinned salmon Canada's third most important export after timber and fur during the late nineteenth and early twentieth centuries. Today, both sport and commercial salmon fishing are major contributors to the BC economy, generating on the order of a $1 billion a year in revenues. In fact, so vital are salmon to the BC economy that a form of seasonal anxiety disorder afflicts the entire province every spring in anticipation of adult salmon returning to the rivers to spawn. The numbers of returning salmon from year to year look something like a heartbeat; they take regular ups and downs every few years. Very literally, the pulse of the province can be read in the fluctuations of salmon population size.

Concern over the health of salmon populations has led to various efforts to supplement wild stocks. Since the 1970s, the federal government's Department of Fisheries and Oceans has run hatcheries along major rivers in BC. Today 18 hatcheries are in operation, and they collectively incubate and release millions of juveniles every year. Current practice is to raise and introduce juveniles in hatcheries along the same reaches where the fish naturally return to spawn. This approach makes good economic sense, since hatcheries avoid paying excessive transportation costs shipping salmon fry up and down the coast into what are often remote spawning grounds. Turns out it also makes good biological sense.

Salmon seem to be exquisitely adapted to the particular local conditions they were raised in, seldom surviving well when transplanted to different watersheds. The survival of nonnative fish relative to native fish declines as the geographic distance among populations increases (Fraser et al. 2011). And of course, salmon are well known for their ability to return to the stream of their birth, probably by using cues such as stream odors to find their way. The incredible site fidelity that salmon display has the effect of reducing gene flow among populations and increasing the opportunity for natural selection to generate local adaptation.

The degree of local adaptation in salmon populations is truly remarkable. Consider this: starting in the late 1800s and not ending until the mid-1980s, efforts to transplant Pacific salmon into otherwise salmon-free reaches of river to establish self-sustaining populations have almost uniformly failed. A comprehensive review of the literature on salmonid transplantation by Withler (1982) highlighted only two successful cases: one was the introduction of Californian chinook salmon to New Zealand in 1901; the other was the introduction of pink, coho, and chinook into the Great Lakes. Both cases involved the

---

1. These names come from Mark Hume and Harvey Thommasen's book, *River of the Angry Moon,* which gives a lyrical account of fly-fishing the Bella Coola River. Published by Greystone Books, an imprint of Douglas and McIntyre Ltd., 1998.

establishment of a new population well outside the native range for American Pacific salmon. Salmon populations can be moved to different reaches in the same river system, especially when a physical barrier such as a rockslide has blocked their way, but transplants of salmon stocks between different watersheds within their native range invariably fail.

The causes of transplant failure remain a mystery. In a few well-documented cases, transplanted fish have succumbed to infection by pathogens (reviewed by Federenko and Shepherd 1986), but any number of other factors could be involved. Water chemistry, nutrient levels, competitors, and predators have all been mentioned as potential culprits. There are even suggestions that the cardinal direction faced by the fish when entering a river from a stream could be important, though it's hard to see how salmon could be such sophisticated practitioners of geomancy or feng shui.

Populations of salmon on the BC coast have had access to their spawning grounds for somewhere in the range of 10,000 to 15,000 years since the retreat of the last glaciers. Ecological specialization to local conditions must therefore have evolved quite quickly. Is what we know about the evolution of local adaptation compatible with this time frame? The geographic scale of local adaptation in salmonids seems to be on the order of hundreds of kilometers, which is about the scale that delimits the boundaries of watersheds along the BC coast. If so, then each watershed represents a distinct environment that imposes strong directional selection. This raises the issue of just how specific adaptation can become to a given set of conditions, and how different environments need to be to support distinct kinds of adaptations. These are some of the topics we consider in this chapter.

There's another more fundamental reason for understanding the evolution of specialization, though. Divergent selection can maintain diversity if and only if different genotypes (or species) are fittest in different environments; that is, if those types are specialized to the prevailing conditions of growth. Although the models we use in population genetics and ecology have taught us a lot about the conditions required to maintain diversity, these models typically assume that a wide range of specialized types are available to selection to begin with. Much less attention has been paid to how specialization evolves and to the fitness trade-offs expected to accompany it. Therefore, in this chapter we focus on how effective divergent selection is at generating specialization and fitness trade-offs across environments.

THE CAUSES OF DIVERGENT SELECTION AND ITS EVOLUTIONARY EFFECTS

An environment that is composed of different patches or habitats offering distinct conditions of growth can be said to be one that varies in space. Spatial variation generates divergent natural selection, in the sense that different optimal phenotypes are favored in different conditions. Provided that little or no gene flow takes place between these two patches, selection should eventually lead to the evolution of the optimal phenotype in each patch. In this section, we focus on how adaptation to one patch affects fitness in other patches.

The outcome of divergent selection is expected to be the evolution of specialization and, so, local adaptation (Felsenstein 1976; Futuyma and Moreno 1988; Kassen 2002). It is important, however, to be absolutely clear about what we mean by these two terms. Specialization, in the ecological sense, occurs when a genotype has high fitness in one or a few sets of patches or habitats that it normally encounters, and low fitness in others. Local adaptation is closely related to the idea of specialization, but it need not be the same thing. It usually means that a genotype or population is specialized to the same habitat where it is normally found in nature. Dispersal or environmental change can therefore dislocate a population from its "normal" habitat, making it specialized but not locally adapted.

Recent reviews of local adaptation in natural populations show that from 60 percent to 80 percent of studies surveyed find some evidence for local adaptation, suggesting that often the two terms can be used interchangeably (Fraser et al. 2011; Greischar and Koskella 2007; Hereford 2009; Hoeksema and Forde 2008; Sanford and Kelly 2011). It's worth noting, however, that the magnitude of fitness differences between the "native" and "foreign" populations can be highly variable: native populations of salmonids were about 20 percent fitter than foreign populations in reciprocal transplant experiments (Fraser et al. 2011), while a survey that included many more species than just salmon suggested that natives outperformed foreigns by 45 percent (Hereford 2009). Interestingly, fitness in a population's own environment is not associated with a dramatic loss of fitness in alternative environments; Hereford (2009) estimates the correlation between "home" and "away" fitness to be just −0.14, certainly not evidence of extreme local adaptation to abiotic conditions. Taken together, these results suggest that local adaptation is commonplace, but the degree of specialization is slight: most genotypes in nature appear to be rather broadly adapted to the usual set of environmental conditions they experience, and specialists are only slightly better adapted to a narrow subset of these conditions.

These observations raise the question of how quickly, and to what extent, specialization evolves under divergent selection. They also raise a second question of whether divergent selection is strong enough to maintain diversity, even when dispersal among environments is low. For diversity to be maintained, different genotypes have to be favored in different environments. As we'll see, this is a rather easy requirement to fulfill even if the genotypes themselves are not specialized to any great extent. These are the two main topics in this chapter. We'll also look at various related issues, including the genetic causes of trade-offs and the contribution made by antagonistic pleiotropy and mutation accumulation to costs of adaptation.

THE CONTRIBUTION OF SINGLE MUTATIONS TO SPECIALIZATION

Specialization evolves because natural selection occurs among genotypes only in the environment where they are found. Natural selection is effectively ignorant of any genotype's fitness in alternative environments. Adaptation therefore happens to the focal environment without regard to fitness in alternative environments.

This idea may seem obvious, but it speaks to an important principle underlying the evolution of specialization. A gene is selected in a particular environment because it is beneficial there. The fitness of that gene in another environment, while unimportant for adaptation to the focal environment, does affect how specialized the evolved genotype will be. Most of the time, a gene that increases fitness in a focal environment does not increase fitness by as much in an alternative environment. It's easy to intuit why. Very often, a kind of functional interference prevents a gene from being beneficial under all conditions. A spoon is useful for eating soup but not for cutting bread, after all. Such functional interference causes *antagonistic* or *negative pleiotropy:* a gene beneficial in one environment reduces fitness in another. Antagonistic pleiotropy is often viewed as the underlying cause of trade-offs in fitness across environments, and so the evolution of specialization occurs.

However, it's important to remember that antagonistic pleiotropy is just one end of a range of pleiotropic fitness effects. Genes that are beneficial in one environment may also be beneficial in other environments, especially if the two environments are very similar. A spoon may not be good at cutting bread, but if dinner is stew, it can be pressed into playing the role of a fork. In this case we could say the gene displays *synergistic* or *positive pleiotropy*. Genes with positive pleiotropic effects can contribute to the evolution of specialization as well. If a gene that is beneficial in the selection environment also increases fitness, though by a lesser amount, in an alternative environment, the result will still be a specialist, since the genotype has higher fitness in the environment of selection than elsewhere. Understanding what determines the pleiotropic effects of beneficial mutations in different environments is vital to understanding the evolution of specialization.

A Theory of Pleiotropy

Pleiotropy is usually described as the manifold effects of a single gene on different phenotypic characters. Most often, these phenotypic characters are thought of as different traits in the same individual. However, Falconer (1981) pointed out that it is equally appropriate to treat the value of the same character in different environments as two characters that are genetically correlated. In a similar vein, we can understand the fitness of the same gene across different environments as being a form of pleiotropy.

A simple verbal model for the direction of pleiotropy across environments is as follows. If two environments are similar in most respects, then a gene beneficial in one is likely to be beneficial in another. On the other hand, if two environments are highly contrasted in their conditions of growth, then a gene beneficial in one environment is unlikely to be beneficial in the other, and it may even be detrimental. The degree to which beneficial fitness effects are "carried over" to different environments should thus depend on how similar the environments are, at least in terms of how they affect expression of the gene of interest. The more highly contrasted the environments, the less likely it is that a gene beneficial in one environment will also be beneficial in the other.

We can formalize this interpretation by extending Fisher's geometric model of adaptation, outlined in Chapter 2, to include multiple phenotypic optima (G. Martin, unpublished). Consider the simplest case of the two-optima model illustrated in **Figure 3.2**, where the optima represent fitness peaks in two different environments.

Mutations that are beneficial in one environment are, like the ones in the simpler one-optimum version of Fisher's model, those that bring the population closer to the optimum in that environment. If these are substituted without regard to their fitness in the alternative environment, we can immediately see that the pleiotropic effect of a mutation beneficial in one environment depends on the starting position of the wild-type genotype relative to the position of the two fitness optima.

When the starting genotype is located between the two peaks (filled circle), most mutations are expected to have antagonistically pleiotropic effects. If

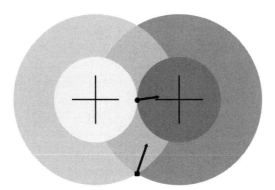

Figure 3.2 Fisher's geometric model with two optima. The pleiotropic effects of mutations that improve adaptation to the dark gray environment (black arrows) are positive if the founding population is located far from the optima in both environments (black square) and negative if it is between them (black circle).

the starting genotype is far away from both peaks (filled square), mutations beneficial in one environment will also be beneficial in the other, although the magnitude of the pleiotropic effect usually will be less than the gain in fitness in the environment of selection. This model can thus account for trade-offs that are underlain by a cost of adaptation as well as those that result from mutations that increase fitness in both the focal and alternative environments. It also predicts that the pleiotropic effects of beneficial mutations are likely to change as a result of adaptation. An initially maladapted genotype that is far from both peaks will tend to generate mutations that are beneficial in both environments. As a population adapts toward one or both peaks, the pleiotropic effects of mutations should tend to decrease, eventually becoming negative.

Data on the Pleiotropic Effects of Beneficial Mutations

There is just one test of the prediction that pleiotropic effects become more antagonistic over the course of adaptation. Jasmin and Zeyl (2013) first isolated genotypes of yeast from a previous selection experiment that had evolved for up to 5000 generations in a glucose medium. Predictably, these genotypes showed a range of adaptation to glucose, those from the start of the experiment having lower fitness than those from later in the experiment. The authors then selected these evolved genotypes on a different resource, galactose, for just over 100 generations. Beneficial mutations in galactose had pleiotropic effects on glucose that depended on how well adapted the founding genotypes were to glucose initially: When the starting genotype was poorly adapted to glucose, mutations that increased fitness on galactose also increased fitness on glucose. However, adaptation to galactose had the pleiotropic effect of lowering fitness on glucose when the genotype was initially well adapted to glucose. Thus, consistent with what the two-optima model predicts, beneficial mutations become increasingly antagonistic as adaptation proceeds.

Two studies have taken a different approach to understanding how specifically adapted beneficial mutations can be by examining directly the pleiotropic effects of beneficial mutations across a range of environments. The first (Bataillon, Zang, and Kassen 2011) assayed the pleiotropic effects of mutations conferring resistance to the quinolone antibiotic nalidixic acid in *Pseudomonas fluorescens*. Resistance to quinolones is known to be conferred by a range of distinct DNA sequence changes in many different genes, the most common being DNA gyrases and topoisomerases that help maintain the supercoiling structure of DNA during replication and membrane-associated efflux pumps that transfer small molecules out of the cell (Jacoby 2005; Poole 2005; Wong and Kassen 2011). The fitness of mutations previously identified as beneficial in a standard permissive medium was measured across a wide range of environments that differed in carbon substrates available for growth.

The fitness of these mutations was almost never deleterious, and some were even modestly beneficial in the novel environments. Similar results were seen in the second study. Ostrowski and colleagues assayed the fitness of the first beneficial mutation fixed in populations of *E. coli* evolving on a glucose minimal medium in alternative carbon sources that differ in the degree to which they share uptake and catabolism pathways with glucose (Ostrowski, Rozen, and Lenski 2005). Their results are reproduced in **Figure 3.3**, where you can immediately see that most mutations were not deleterious in the alternative environments. It is worth noting that in neither study was there a tendency for mutations of large beneficial effect to be more deleterious than those of more modest effect.

We can get some further insight into the pleiotropic effects of single mutations from the literature on the effects of antibiotic resistance mutations in the absence of drugs. Resistance mutations are an extreme kind of beneficial mutation, since they make the difference between life and death when a drug is present. Drugs typically target components of cellular life that are essential for growth—cell wall synthesis and assembly (beta-lactams), DNA supercoiling and replication (quinolones, mentioned earlier), and RNA transcription

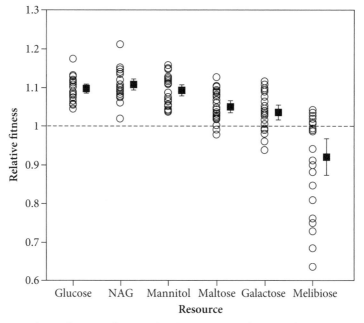

Figure 3.3 Relative fitness in five novel carbon sources of 27 *E. coli* mutants that were among the first to fix during adaptation to a glucose. Each circle is the mean of three replicate fitness measures, except in glucose where 15 replicate measures were taken. Filled squares are the means of all mutants, and error bars represent 95 percent confidence intervals for *n* = 27. Reproduced with permission from Ostrowski, Rozen, and Lenski (2005) © 2007, John Wiley and Sons.

and translation (aminoglycosides)—so a mutation that confers resistance is expected to be costly in the absence of drugs. However, this is not always the case. A recent review of costs of resistance among common resistance mutations in pathogenic microbes suggests that costs can be highly variable and dependent on the environment and genetic background in which they are tested (Andersson and Hughes 2010). This finding could explain why some resistance mutations are common in clinical settings: if they confer resistance and pay little in the way of costs, these mutations would be favored in the presence of drug and not be at a selective disadvantage in the absence of drug. Such "no-cost" mutations would thus be commonly isolated from patients experiencing repeated rounds of antibiotic therapy.

These examples might give the impression that antagonistic pleiotropy is not often observed in microbial experiments. However, antagonistic pleiotropy has certainly contributed to the evolution of specialization in some microbial experiments. The highly repeatable diversification that occurs in static microcosms of *Pseudomonas fluorescens* mentioned in Chapter 1 is one of the most compelling examples. The ecological transition from broth to air–broth interface is caused by a range of mutations that allow biofilm formation, and so colonization of the air–broth interface, at the cost of low fitness in the broth phase (Bantinaki et al. 2007; Rainey and Travisano 1998; Rainey and Rainey 2003). Other examples include host expansion in phage (Duffy, Turner, and Burch 2006), adaptation to minimal media with a single carbon source (Conrad et al. 2010), and adaptation to different carbon sources (Zhong et al. 2004), to highlight just a few.

The take-home message from these studies is that antagonistic pleiotropy is not by any means a general or consistent property of beneficial mutations. Rather, the degree of pleiotropy is a property of a mutation that depends on the ecological and genetic context in which it is measured. The two-optima version of Fisher's model presented earlier suggests it may be a property that evolves in a predictable way. More direct tests of this idea are welcome.

Pleiotropic Effects of Neutral Mutations

The two-optima version of Fisher's model is useful for understanding the range of fitness effects associated with mutations under selection, but what about neutral mutations? Because the fixation rate of neutral mutations is just their mutation rate, they are unlikely to be major contributors to specialization on the relatively short time scales of most experiments. However, most microbial selection experiments involve asexual organisms, so neutral or even mildly deleterious mutations may hitchhike to high frequency in a population because they are linked to a beneficial mutation. The pleiotropic effects of these hitchhiking mutations cannot be ignored.

The pleiotropic effects of neutral mutations are expected to be either neutral or deleterious in alternative environments, depending on where in the genome the mutation occurs. A mutation that occurs in a genomic region that is not expressed or is not otherwise important in gene expression in a given environment is likely to be neutral in that environment. On the other hand, a mutation that's neutral in one environment may become deleterious in a novel environment if the genomic region where it occurs is now expressed in the novel conditions. The mutation is likely to be deleterious because a random change in a gene is far more likely to disrupt, rather than improve, function. No studies have examined experimentally the range of pleiotropic fitness effects among neutral mutations substituted during a selection experiment, although some studies have suggested that they are important contributors to the evolution of trade-offs (MacLean and Bell 2002; Reboud and Bell 1997).

It's worth mentioning that the extent to which any region of the genome is genuinely neutral is a matter of some debate (ENCODE Project Consortium 2012; Kapranov and St. Laurent 2012). Even the canonical example of neutral mutations, synonymous substitutions that do not change the amino acid sequence of the protein, can be under selection (Cuevas, Domingo-Calap, and Sanjuán 2012; Parmley and Hurst 2007; Plotkin and Kudla 2011). The available evidence suggests that the fitness effect of synonymous mutations under selection is typically deleterious, although a recent experiment has uncovered two synonymous mutations in the same gene (a transporter of glucose in Pseudomonas fluorescens) that are not only clearly beneficial, but increase fitness by an amount (~6–9%) comparable to any non-synonymous mutation (Bailey et al. in review).

When it comes to understanding the evolution of specialization, however, the argument about the pervasiveness of neutrality across the genome is a bit of a red herring. What matters, instead, is whether or not a given mutation, arising under a particular set of conditions, behaves as if it is neutral. Since neutral mutations commonly occur and can be fixed (Eyre-Walker and Keightley 2007; Hietpas, Jensen, and Bolon 2011), they cannot be ignored for their contribution to the evolution of specialization.

MULTIPLE MUTATIONS

So far, we have considered the pleiotropic fitness effects of a single beneficial mutation in novel environments. What happens when selection is allowed to proceed for sufficiently long periods of time that multiple beneficial mutations arise and fix sequentially? Because these genes occur in a common genetic background, we need to consider their combined pleiotropic effects on fitness in an alternative environment when evaluating their impact on specialization.

In other words, we need to look at the epistatic fitness effects of multiple mutations when assayed in environments where they were not selected.

Just as the pleiotropic effects of a single mutation are expected to become more negative as adaptation proceeds, so too should the combined pleiotropic effects of multiple mutations. Trade-offs in fitness across environments thus evolve thanks to the substitution of multiple mutations. The proximate reason for trade-offs is that the fitness gain associated with a mutation in a focal environment is nearly always larger than it is in an alternative environment. This happens because, as explained earlier, the improvement in function that occurs in the focal environment is unlikely to be carried over to the same extent in an alternative environment, and may even be antagonistically pleiotropic there. As selection fixes more mutations in the selection environment, the contrast in fitness across environments is expected to increase. Over time, conditionally neutral mutations will also accumulate, contributing to an ever-narrowing degree of specialization. When a lineage becomes so specialized that it actually has a lower fitness in an alternative environment than the ancestor it was derived from, we say that specialization has evolved a cost of adaptation.

We can detect trade-offs in fitness associated with specialization in a straightforward assay called a reciprocal transplant. Basically, we measure the fitness of a pair of specialist genotypes or populations in the environments where they had evolved as well as in the alternative environment where their sister lineage had evolved. Specialization is evidenced by a negative genetic correlation in fitness across those environments. That is, when we draw a graph with fitness in a pair of environments along the axes, the slope of the line connecting the fitness of the two genotypes in each environment should be negative.

It's important to remember that such trade-offs, and the negative genetic correlations that result, can evolve in two fundamentally different ways. To see this, consider **Figure 3.4**, which illustrates the consequences of selection in a focal environment (black arrow) on fitness in an alternative environment (gray arrow). The top two panels illustrate a situation where adaptation to one environment leads to a loss of fitness, relative to the ancestral genotype, in the alternative environment. In this situation, specialization is underlain by a cost of adaptation. The result, in terms of the genetic correlation in fitness, can be seen in the last panel on the right: it is negative, indicating a fitness trade-off.

The bottom two panels of the figure show a scenario where there is no cost of adaptation, but a trade-off in fitness still evolves. Fitness increases in both the focal and alternative environments, but the gain in fitness is not as large in the latter as in the former. In both cases the result is the evolution of specialization, in the sense that each type is best adapted to the environment where it was selected, and a negative genetic correlation in fitness. Where selection is divergent, and we are considering adaptation to a pair of environments,

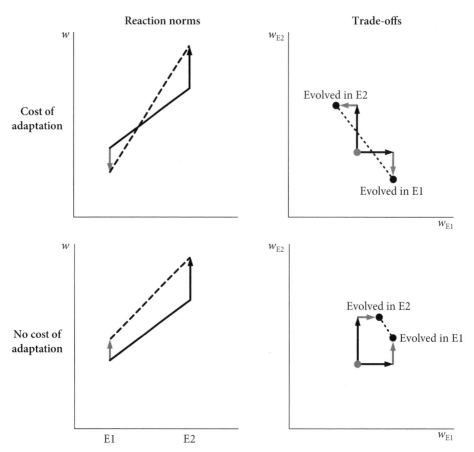

Figure 3.4 The evolution of fitness trade-offs across environments. Left-hand panels depict the response of absolute fitness (w) to selection in E2, a high-quality environment (black arrows), and a poor-quality environment (E1; gray arrows). Lines connect the fitness of a genotype in two environments (called a reaction norm), the ancestor represented by a solid line and evolved genotypes by dashed lines. The right-hand panels depict the fitness of each evolved genotype (black circles) and the ancestor (gray circle) in both environments. Note that a trade-off in fitness (thin dotted line) evolves in both the presence (top right) and absence (bottom right) of costs of adaptation.

the emergence of trade-offs generates a negative genetic correlation in fitness across the two environments between the two evolved genotypes.

The Combined Pleiotropic Effects of Multiple Mutations

Direct tests of the prediction that combinations of beneficial mutations tend to lead to less positive or even negative fitness effects in alternative environments have not been done. In antibiotic resistance studies, researchers occasionally observe that costs of resistance increase with the number of mutations involved in confering resistance. But, as is the case for single resistance mutations, there is substantial variability depending on the drug, species, and environments being considered (Andersson and Hughes 2010). A recent study has tracked the dynamics of how specialization evolves to different carbon substrates in Pseudomonas fluorescens over 1000 generations, where we know that multiple mutations have been substituted (Bailey and Kassen 2012). Specialization increased in all lines by the end of the experiment, as expected. Specific genetic causes of the increase in specialization—whether it was due to the eventual substitution of rare, large-effect mutations or caused by the combined effects of many smaller mutations—have yet to be worked out. In general, we lack a clear understanding of how the pleiotropic effects of multiple mutations evolve. This is an open avenue for further research.

The Evolution of Fitness Trade-offs and Negative Genetic Correlations

How readily do trade-offs evolve in laboratory selection experiments? To answer this question, let's look at data collated from the literature on the results of divergent selection from an initially isogenic ancestor in bacteria

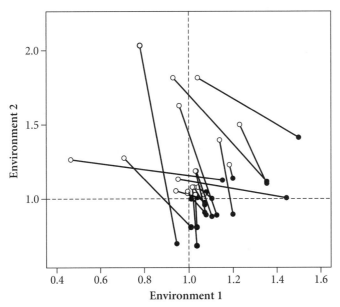

Figure 3.5 Trade-offs in fitness resulting from divergent selection. Lines connect a pair of evolved populations descended from a common ancestor selected in one of two environments (filled circles: selected in environment 1; open circles: selected in environment 2). Fitness estimates come from full reciprocal transplant experiments, and the fitness of each population is measured in both environments. Because values are expressed as relative fitness, the fitness of the ancestor is by definition given by the intersection of the two dashed lines in the figure.

Table 3.1 Fitness of evolved lines following divergent selection in a focal (environment of selection) and alternative environments.

Species	Generations	Citation	Focal environment	Alternative environment	Fitness in focal environment (w_f)	Fitness in alternative environment (w_{alt})	Standard error of w_f	Standard error of w_{alt}
BACTERIA								
Escherichia coli	1000	Hughes, Cullum, and Bennett 2007	basic	acidic	1.01	0.809	0.033	0.062
			acidic	basic	1.276	0.708	0.144	0.03
Escherichia coli	2000	Bennett, Lenski, and Mittler 1992	32°C	37°C	1.104	1.041	0.028	0.035
			32°C	42°C	1.104	0.881	0.028	0.103
			37°C	32°C	1.078	1.002	0.019	0.015
			37°C	42°C	1.078	0.892	0.019	0.083
			42°C	32°C	1.188	1.034	0.058	0.028
			42°C	37°C	1.188	1.03	0.058	0.023
Escherichia coli	2000	Cooper and Lenski 2010	glucose	maltose	1.199	1.173	0.016	0.021
			maltose	glucose	1.272	1.183	0.012	0.017
			glucose	lactose	1.199	0.92	0.016	0.012
			lactose	glucose	1.45	1.139	0.031	0.026
Methylobacterium extorquens	1500	Lee, Chou, and Marx 2009	methanol	succinate	1.145	1.126	0.01	0.019
			succinate	methanol	1.26	1.143	0.008	0.005
Pseudomonas fluorescens	600	Jasmin and Kassen 2007a	glucose	mannose	1.072	0.908	0.024	0.019
			glucose	mannitol	1.072	0.959	0.024	0.007
			glucose	sorbitol	1.072	0.975	0.024	0.003
			mannose	glucose	1.037	0.995	0.020	0.013
			mannose	mannitol	1.037	0.683	0.020	0.053
			mannose	sorbitol	1.037	0.809	0.020	0.033
			mannitol	glucose	1.078	1.023	0.046	0.002
			mannitol	mannose	1.078	1.016	0.046	0.017
			mannitol	sorbitol	1.078	1.046	0.046	0.026

Species	Generations	Citation	Focal environment	Alternative environment	Fitness in focal environment (w_f)	Fitness in alternative environment (w_{alt})	Standard error of w_f	Standard error of w_{alt}
Pseudomonas fluorescens	500	Jasmin and Kassen 2007b	sorbitol	glucose	1.028	1.027	0.012	0.009
			sorbitol	mannose	1.028	1.029	0.012	0.011
			sorbitol	mannitol	1.028	1.069	0.012	0.030
			mannose	xylose	1.126	0.890	0.010	0.140
			xylose	mannose	1.629	0.957	0.065	0.011
Pseudomonas fluorescens	1000	Bailey and Kassen 2012	glucose	mannose	1.353	1.103	0.101	0.063
			glucose	xylose	1.353	1.117	0.101	0.035
			mannose	glucose	1.5	1.23	0.082	0.067
			mannose	xylose	1.5	1.413	0.082	0.137
			xylose	glucose	1.817	0.93	0.108	0.061
			xylose	mannose	1.817	1.04	0.108	0.064
EUKARYOTES								
Chlamydomonas reinhardtii	500	Bell and Reboud 1997	light	dark	0.947	0.698	0.05	0.184
			dark	light	2.034	0.781	0.783	0.06
Saccharomyces cerevisiae	500	Dettman et al. 2007	salt	glucose	1.444	1.006	0.069	0.011
			glucose	salt	1.132	0.952	0.038	0.019
Neurospora crassa	1500	Dettman, Anderson, and Kohn 2008	salt	temp	1.038	1.01	0.015	0.02
			temp	salt	1.053	0.942	0.017	0.036

Note: Standard error reported here is a measure of dispersion among replicate selection lines.

and fungi. The data includes only those experiments that provided measures of fitness obtained from a full reciprocal transplant of the evolved lines in the alternative environments. **Figure 3.5** summarizes the results, details of which are presented in **Table 3.1**. In Figure 3.5, the means among replicate selection lines were calculated, and so the actual number of independent instances of evolution is actually greater than that shown (measures of dispersion are not shown, to avoid making the figure too cluttered to interpret).

Divergent selection generated negative genetic correlations in fitness across environments in 20 of 21 cases. This is highly significant in a binomial test, where there is an equal probability of getting a positive or negative correlation ($P < .0001$). The one positive correlation, which cannot really be seen because it's hidden in the mess of points that are clustered close to the origin, was in a study that looked at adaptation to mannitol and sorbitol in *Pseudomonas fluorescens* (Jasmin and Kassen 2007a). The mechanistic reason that a trade-off didn't evolve in this instance isn't known, but is probably because mannitol and sorbitol were taken up and metabolized by the same pathway; thus, from the perspective of the strain that was used, they are effectively the same resource.

The take-home point of this figure is that trade-offs readily evolve, often quickly. The duration of the experiments included in the figure varied from a few hundred to a few thousand generations, well within the time frame within which, for example, salmon populations in BC are thought to have evolved local adaptation.

Costs of Adaptation

How often are trade-offs the result of a cost of adaptation, meaning that adaptation to one environment caused a loss of fitness relative to the ancestor in the alternative environment? This happened in 16 of the 21 experiments plotted in Figure 3.5. In five of those 16 experiments, the cost was symmetric in the sense that divergent selection generated costs of adaptation in both environments. In the remaining 11 experiments, only one line showed evidence for a cost of adaptation while the others increased in fitness in both environments. In five additional experiments, there was absolutely no evidence of a cost of adaptation in either pair of lines. Trade-offs in fitness evolved in these experiments because fitness improved more in the environment of selection than it did in the alternative environment. The data from selection experiments in microbes thus provides examples of trade-offs evolving both in the presence and absence of costs of adaptation.

As explained earlier, when a cost of adaptation evolves, it may be due to substitution of mutations with antagonistically pleiotropic effects, the accumulation of mutations that are neutral in the environment of selection but deleterious elsewhere, or some combination of the two. Which of antagonistic pleiotropy or mutation accumulation makes the largest contribution to costs of adaptation in selection experiments? The evidence is mixed. Some experiments favor antagonistic pleiotropy and others mutation accumulation (reviewed in Kassen 2002). Still others provide circumstantial evidence to suggest both factors are at play (MacLean and Bell 2002; Maughan et al. 2006; Presloid et al. 2008).

We can get a more comprehensive view by analyzing the data shown in Table 3.1 and by performing a regression of the amount of adaptation in a focal environment against that in an alternative environment. We can decompose the relative contributions of antagonistic pleiotropy and mutation accumulation by examining the slope and intercept of this regression (MacLean and Bell 2002). The logic is as follows: Neutral mutations accumulate through drift, and the rate at which this happens is, according to theory, just the mutation rate; it is independent of the magnitude of the adaptive response in the focal environment. The signature of a cost of adaptation attributable to neutral mutations will therefore be a decrease in the intercept by an amount proportional to the mutation rate. Large advances in the focal environment will be due to the accumulation of beneficial mutations, and if these mutations generally tend to have antagonistic effects in other environments, then the slope of the regression will be negative.

Figure 3.6 shows such a regression for those experiments from Table 3.1 where there was evidence of a cost of adaptation. The slope of the

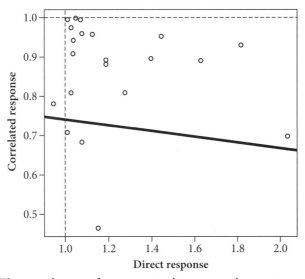

Figure 3.6 The contribution of antagonistic pleiotropy and mutation accumulation to costs of adaptation. Direct responses are a measure of adaptation in a focal environment; correlated response is fitness in an alternative environment (see explanation and interpretation in the text). Note that the regression was calculated by weighting observations by the pooled standard deviation calculated from the variance among replicate lines in each experiment.

regression is negative ($b \pm 1$ standard error $= -.073 \pm .072$) but not significant ($t_{1,19} = -1.020$, $P = .321$), suggesting that antagonistic pleiotropy makes little contribution to costs of adaptation in these experiments. The intercept, on the other hand, is significantly less than one ($a \pm 1$ standard error $= .186 \pm .049$; $t_{1,19} = -3.802$, $P = .001$). The implication is that mutation accumulation is the major source of costs in these experiments.

It does seem odd that mutation accumulation should play such an important role in contributing to costs of adaptation in these experiments. Most selection experiments last from hundreds to thousands of generations but the time needed for fixation of a neutral mutation is at least on the order of millions of generations, if not substantially longer. There is simply insufficient time for many neutral mutations to accumulate in microbial selection experiments.

It seems more likely, given the large population sizes in microbial experiments, that mutation supply rates would be high enough that neutral or mildly deleterious mutations are introduced regularly into the population and sweep to fixation along with the beneficial mutations being selected. Such hitchhiking is helped by the fact that reproduction is asexual, and so mutations occurring in the same genome cannot be separated from each other. We can actually get a rough idea of how high the mutation rate is in these experiments if we assume, as the data tells us, that antagonistic pleiotropy contributes little to the cost of adaptation. The loss of fitness in the alternative environment, w_{alt}, due to mutation accumulation will be a function of the genome-wide mutation rate and time. In other words, $w_{\text{alt}} = -Ut$, where U is the genome-wide mutation rate and t is time in generations. Table 3.1 provides estimates of w_{alt} and t; therefore, we can estimate U. The result is a genome-wide neutral mutation rate of approximately 2.77×10^{-4} mutations per generation. This value is an order of magnitude less than the "typical" genome-wide mutation rate of about 3×10^{-3} given by Drake (1991), and it's smaller still than more recent estimates based on high-throughput genome-sequencing techniques (Lynch 2010). Mutation rates are thus probably high enough for a substantial number of neutral mutations to accumulate by hitchhiking in most laboratory selection experiments.

We shouldn't be too quick to dismiss the role of antagonistic pleiotropy as a cause of costs of adaptation in these experiments, however. For one thing, the analysis assumes a linear decline in the fitness in the alternative environment when, in fact, it could take any shape at all. We simply don't know enough about how the pleiotropic effects of multiple beneficial mutations combine to be able to say much more. A second reason is that only one, or at most two, alternative environments were used in these experiments. Other experiments not included here (because they don't fit the criterion of having examined divergent selection) have detected antagonistic pleiotropy by examining correlated responses in a much wider range of alternative environments (Cooper and Lenski 2000; MacLean, Bell, Rainey 2004; Travisano and Lenski 1996) or have examined the decay of a particular function or trait known a priori to be not under selection (Dettman, Anderson, and Kohn 2008; Maughan et al. 2006). Evidence for antagonistic pleiotropy can be obtained by examining the dynamics of how costs evolve—these should track increases in fitness in the selected environment—and how closely the functional losses across a range of substrates parallel each other. Thus this data set may be biased toward detecting costs due to mutation accumulation.

Asymmetries in the Response to Divergent Selection

One of the more notable results from Figure 3.5 is that many divergent selection experiments show responses to selection in alternative environments that are asymmetric. In other words, adaptation to environment 1 may result in a cost of adaptation to environment 2, but adaptation to environment 2 leads to an increase in fitness in environment 1. Why might this be?

Quantitative geneticists have recognized for some time that responses to selection can sometimes be asymmetric. But the explanations they have come up with—drift, inbreeding depression, maternal effects, changes in gene frequency that change heritabilities (Bohren, Hill, and Robertson 1966; Villanueva and Kennedy 1992)— typically involve selection from standing genetic variation in diploid populations. These explanations cannot, therefore, be used to understand asymmetric responses when selection operates on de novo variation introduced by mutation in haploid populations, as in most microbial selection experiments. We must look elsewhere for explanations.

One possibility is that the selected genes have drastically different pleiotropic fitness effects in their respective alternative environments. Zhong et al. (2004), for example, reported that selection of *E. coli* in chemostats containing either lactulose or methyl-galactosidase as the sole carbon source resulted in specialization caused by mutations unique to each resource. Selection on lactulose tended to lead to adaptation through duplications of the *lac* operon and had the pleiotropic effect of increasing fitness on methyl-galactosidase. However, selection on methyl-galactosidase is never achieved through *lac* duplications but instead occurs through mutations in either *galS*, which cause constitutive expression of an ATP-binding cassette transporter specific to methyl-galactosidase, or *mgl*, which forces transport through the *lac* permease. At least in the case of *galS* mutations, there appears to be a strongly antagonistic effect on fitness in lactulose.

A second explanation is that there are differences in the number of genes responsible for adaptation. This situation can generate asymmetries in fitness in alternative environments in the following way. Suppose that more genes are expressed in one environment than the other, and those expressed in the latter are a subset of those in the former. Asymmetry may result because mutation accumulation can potentially disrupt function in a larger range of targets when adaptation occurs in the environment where fewer genes are expressed (Lee, Chou, and Marx 2009). It's also possible that having a larger target can lead to a higher effective mutation supply rate, which should favor the substitution of genes of larger effect. If these genes have any tendency to be more antagonistically pleiotropic than genes of smaller effect—a pattern that currently has little empirical support—this could further contribute to asymmetric correlated responses.

CONCLUSIONS

Salmon populations along the Pacific coast of North America must be among the most striking and compelling examples of rapid local adaptation known. They exhibit such extreme specialization to natal spawning sites that attempts to transplant fish from one watershed to another to establish self-sustaining populations over the past 100 years have uniformly failed. Divergent selection would need to have been outstandingly strong for such precise specialization to evolve as quickly as it did, within the last 10,000 to 15,000 years since the glacial ice sheets retreated. Given the commercial importance of salmon to Pacific coast economies, it's surprising that more is not known about causes of such precise local adaptation (Primmer 2011).

The ideas and data reviewed in this chapter take us some way toward understanding how such incredible specialization could have evolved so fast. The key points are summarized here.

1. A single beneficial mutation substituted in one environment can have pleiotropic effects on fitness in a novel environment that ranges from modestly positive to strongly negative. Almost invariably, fitness increases more in the environment of selection than it does in an alternative environment, even if the pleiotropic fitness effect in the alternative environment is positive.

2. A simple two-optima extension of Fisher's geometric model can in principle account for the variation in pleiotropic fitness effects across environments. The geometry of adaptation in this model confirms a common experimental observation: the pleiotropic fitness effect of a mutation substituted in one environment is almost always less than the gain in fitness in the environment of selection.

3. Because fitness gains in selection environments exceed those in alternative environments, divergent selection generates fitness trade-offs that lead to a negative genetic correlation in fitness across environments. Sometimes these

trade-offs are underlain by a cost of adaptation, meaning a loss of fitness in an alternative environment (relative to the ancestor), but they need not be.

4. The genetic causes of costs of adaptation can sometimes involve antagonistically pleiotropic fitness effects, but the available data from microbial selection experiments points to mutation accumulation as playing a more important role, likely because neutral mutations hitchhike to high frequency alongside a beneficial mutation due to the lack of recombination in most microbial experiments. These neutral mutations are actually conditionally neutral, in the sense that they become deleterious when expressed in a novel environment.

5. Divergent selection may often generate asymmetric responses to selection when pairs of lineages specialized to their own environment are tested in each other's environments. The causes of asymmetric responses have not been explored in detail but could be due to differences in the pleiotropic fitness effects of the genes substituted or differences in the range of genetic targets under selection.

The experiments reviewed here document the emergence of specialization and trade-offs on time scales ranging from a few hundred to a few thousands of generations. Salmonids on the Pacific coast would have had access to ice-free spawning grounds starting about (conservatively) 10,000 years ago. The average generation time for salmonids can be between two and seven years, so if we take our most conservative estimate of the number of generations available for divergent selection to do its work, this puts the time frame for the evolution of local adaptation in salmonids at about 1400 generations. This calculation assumes, of course, that there is no dispersal or gene flow, which almost certainly has happened. Nevertheless, if—and currently this is a big if, because we know little about the genetics of adaptation in salmonids—divergence also was helped along by the substitution of genes with strongly antagonistic fitness effects, it would not be hard to see how such extreme specialization could evolve so quickly.

It remains, of course, something of a leap of faith to draw a close parallel between the evolution of local adaptation in salmonids and that in microbial populations evolving in test tubes. The latter, for one reason, no doubt have much larger population sizes than the former. Microbes in test tubes also lack recombination, and this may help explain why conditionally neutral mutations seem to be so important in contributing to costs of adaptation in these experiments. Of course, such conditionally neutral mutations may also be fairly common in salmonids, although for a different reason: effective population sizes may be sufficiently small that neutral or mildly deleterious mutations occasionally become substituted. How important conditionally neutral mutations are in contributing to costs of adaptation, as opposed to genes with strongly antagonistic fitness effects across environments, remains an important avenue for future research. Their relative contributions may be predictable to some extent: antagonistic pleiotropy should be important when environments are strongly contrasted, whereas mutation accumulation should contribute more to costs of adaptation as time goes on. These predictions await strong experimental tests.

Understanding the nature of pleiotropy and how it contributes to adaptation remains one of the biggest gaps in evolutionary knowledge. The two-optima extension of Fisher's geometric model is one attempt to fill this gap, and it's appealing because it seems to account for results commonly observed in experimental evolution. Many selection experiments in microbes find that adaptation to one set of conditions often results in adaptation to others used in the lab. If most selection experiments are founded with genotypes that are maladapted to laboratory conditions, or at least the conditions used to study divergent selection, then such a result is exactly what we expect under the two-optima model. Most beneficial mutations will have positive pleiotropic effects when the starting genotype is far from both optima. To the extent that this situation mimics what happens during adaptation to novel environmental conditions in nature, then we might also expect that the first mutations substituted should not exhibit strongly pleiotropic effects.

There is also no doubt some room here for more mechanistic studies of pleiotropy. Knowing the mechanistic causes of adaptation to one environment and its consequences for adaptation to alternative environments can be extremely important in some applied situations. Take the causes of resistance to quinolone antibiotics as an example. Quinolones are a unique class of antibiotic, first discovered in 1962 and introduced for clinical use in 1967 as a treatment for urinary tract infections caused by enterococcal bacteria (Emmerson and Jones 2003). They were designed to target enzymes involved in winding and unwinding DNA during replication. Resistance can result from mutations in these enzymes, DNA gyrase or topoisomerase to be specific, that change their conformation in such a way that the drug cannot interfere with replication. Resistance can also be caused by mutations that decrease the drug concentration in the cell, often by constitutive expression of efflux pumps that transport small molecules from the cytoplasm into the environment. Resistance mutations in DNA supercoiling enzymes are very specific: they confer resistance to quinolones but not to other classes of antibiotics like aminoglycosides (streptomycin, for example) or tetracyclines. The same is not true for efflux pump mutations, because they increase the rate at which small molecules including antibiotics are pumped out of the cell. As a result, resistance mediated by efflux pump mutations has the pleiotropic effect of conferring resistance to other classes of drug as well. Making an informed decision on treatment options for a patient infected with a quinolone-resistant pathogen thus depends intimately on the genetic causes of resistance and its pleiotropic effects (Wong and Kassen 2011).

A final gap in our knowledge concerns the importance of ecological interactions—such as those due to competition, predation, or pathogens—in driving specialization and diversification. The work reviewed here has clearly shown that when adaptation occurs to abiotic features of the environment such as temperature or different resources, trade-offs evolve readily. Only sometimes does this lead to the sort of extreme specialization, underlain by large costs of adaptation, that characterizes salmonid populations on the Pacific coast. It's possible that one of the missing factors here in explaining extreme divergence and local adaptation is the additional selective pressures associated with species interactions and the biotic environment more generally. In Chapter 6, we'll take up this subject in more detail.

CHAPTER 4

Selection in Variable Environments

Sometime in 1977 or 1978, a passenger from Europe stepped off a plane in Australia and visited a wheat field. Clinging to the visitor's clothes was a yellowish-brown powder, the spores of the stripe rust, *Puccinia striiformis*. Stripe rust was well known as a major pathogen of wheat in temperate climates, but it had never been found in Australia. Temperatures in summer were much higher and precipitation much less in Australia compared to the wheat-growing regions of Europe, so no one in Australia was concerned that stripe rust would present much of a problem to wheat crops there.

Concern over a stripe rust invasion of Australian wheat began in 1979 with the discovery of the first infected crop. Within a couple of years, a major outbreak was under way. Variants of the original pathogen genotype from 1979 have been recovered every year since (Wellings and Mcintosh 1987; Wellings 2007). The cost to the Australian wheat industry, in terms of prevention and lost production, has been estimated at nearly AUS $1 billion for the year 2010 alone (Murray and Brennan 2009).

Given the obvious economic importance of wheat, stripe rust infections have been monitored closely in Australia ever since it was first discovered. Rust populations are predominantly clonal, and most of the novel variants detected are genetically very similar to preexisting genotypes. Most of the derived strains of stripe rust in Australia thus appear to stem from one, or at most a few, mutations. These mutations, moreover, are usually associated with increased virulence against the most prevalent wheat variety in cultivation.

Figure 4.1 gives some impression of how the abundance of the major rust genotypes across the entire country has changed since 1979. If you squint, you can just about make out a pattern resembling the regular replacement of strains we might expect during adaptation. 104 E137 A–, the strain first detected, was most abundant initially but 10 years later was replaced by 104 E137 A+. By 2001, the strain 134 E16 A+ was on the rise and became the

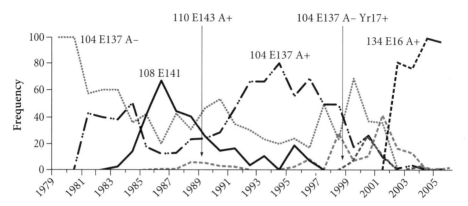

Figure 4.1 Relative frequencies of the major wheat rust pathotypes in Australia between 1979 and 2005. Redrawn from Wellings (2007) *Australian Journal of Agricultural Research* 58: 567-575, http://www.publish.csiro.au/nid/40/paper/AR07130.htm with permission from CSIRO Publishing.

dominant strain in 2005. Admittedly, beyond these two examples it's hard to detect a regular pattern of sequential substitution. In fact, for prolonged periods, multiple strains coexisted simultaneously. You can see this most clearly between 1984 and 1991, and again between 1988 and 2001.

These patterns are at odds with what theory predicts should happen during adaptation to a novel environment. Natural selection should lead to the regular replacement of one genotype for another, and diversity should be present only during the transition periods. As the population becomes better adapted, the supply of new genetic variants increasing fitness should run out, and the population should remain fixed for a single dominant genotype for long periods of time. This is not at all what the stripe rust data reveals: high diversity among pathogen genotypes is more the norm than the exception, and there is little indication that the population shows any tendency to become genetically homogeneous as time goes on. Evidently, adaptation of stripe rust to wheat in Australia is a more complicated process than the one described so far in this book.

These complications may arise because stripe rust probably faces a much more heterogeneous environment than the ones we have considered so far. Wheat is selectively bred for resistance to rust, and so the wheat variety planted can impose a major selective pressure on the rust population. The ability to survive the harsh summer weather is another source of selection, as is climate variability. Moreover, in Australia wheat is planted widely—across the entire width of the country at its southern end, and up the eastern coast to the north. A rust strain could easily "hide out" in a habitat or microclimate that provides a refuge during a particularly harsh summer, reappearing later when conditions are favorable.

The periods of diversity seen in Figure 4.1 may be caused ultimately by selection in a variable environment, so no one rust genotype would be superior across all conditions. This interpretation gains some support from models in population genetics and community ecology that have shown that environmental variation, especially in space, can be an important mechanism supporting both genetic variation in populations and species diversity in communities (Felsenstein 1976; Hedrick 1986; Levins 1968; Morin 2011; Tilman 1982; Vellend and Geber 2005). Environmental variation may not be the only explanation, though. For rust the environment may be quite homogeneous, since wheat is its preferred and most common host. Wheat is grown in fields that are distributed across the country, subdividing the rust population in space. Beneficial mutations arising independently in different subpopulations will then compete, generating periods of apparent coexistence among rust genotypes across the country as a whole.

Which of these two explanations—environmental variation favoring different genotypes in different regions, or environmental homogeneity with a spatially subdivided population—best explains the dynamics of rust genotypes across Australia? It's not something that can be decided at this point. There are, as yet, no clear predictions as to how either mechanism changes the dynamics of genotype frequencies on the sorts of time scales that would allow us to distinguish between the alternative explanations. Moreover, the crucial pieces of information that could help investigators decide—measures of population structure, migration rate, and how specialized the genotypes are to different environments—need to be collected on such fine scales of time and space that it's probably not economically feasible to do so.

Such endpoints are routine to gather in microbial experiments, though, so there is a real opportunity here for experimental evolution to make a contribution to a very practical problem. Moreover, there's a compelling economic incentive: if scientists can make better predictions—even marginally better ones than we can now—about which pathogen genotype is likely to prevail and when, this will be an immense benefit to the farmers, who have to decide every year about which wheat cultivar to plant. The consequences of making the wrong decision, as we have seen from the price tag associated with wheat rust in Australia, can be devastating.

This chapter summarizes what is known from experimental evolution about how environmental variation affects the dynamics of adaptive evolution and the ecological properties of the genotypes that evolve. Does environmental variation change the predictions of the genetic theory of adaptation to simple, unchanging environments discussed in Chapter 2? If so, how? Under what conditions do we see the evolution of narrowly adapted specialists versus more broadly adapted generalists? When specialists evolve, does selection tend to lead to the evolution of just a single specialized type or can it support multiple specialists?

These questions, and especially the last two, are of direct relevance to the theme of this book. Environmental variation plays a central role in theories for the evolution and maintenance of diversity because it generates selection on traits associated with niche breadth—those that help determine the range of conditions under which an organism can live and reproduce. It also acts to support diversity when distinct niche specialist types are present. Thus it has often been suggested that environmental variation represents a very general explanation for many patterns of diversity in nature, from the evolution of genotypic niche breadth up to patterns of species diversity across landscapes (Levins 1968; Lewontin 1974; Rosenzweig 1995; Tilman 1982; Vellend and Geber 2005). The work reviewed in this chapter takes us some way toward evaluating this claim.

THE NATURE OF ENVIRONMENTAL VARIATION

Most natural environments, and probably many managed ones like crop fields, are highly variable, at least from the perspective of the organisms that occupy them. To understand how adaptation proceeds in a variable environment, let's first look at the different ways that environments can vary.

We can see a variable environment as being composed of a collection of patches or habitats, each with different phenotypic optima. Environmental variation changes the direction of selection in ways that depend on how the different optima are distributed in time, space, and scale. Scale here refers to how frequently a lineage encounters different optima, and it is measured relative to the lifetime of an average individual. If individuals experience many different kinds of patches within their lifetime, we say that the scale of variation is fine-grained. An individual that typically experiences only one kind of patch during its lifetime, but whose offspring may experience a different patch, experiences variation that is coarse-grained. Environmental variation in time generates fluctuating selection; in space, it causes divergent selection. This leads to a classification of different kinds of environmental variation, as illustrated in **Figure 4.2**.

We can further decompose environmental variation into two parts. One part is due to variance among patches in productivity, the number of individuals contributed by a patch to the total population; the other is due to the contrast among patches in the optimal phenotype (**Figure 4.3**). Variance in productivity means that some patches support many more individuals than others. Patches that are highly contrasted have very distinct phenotypic

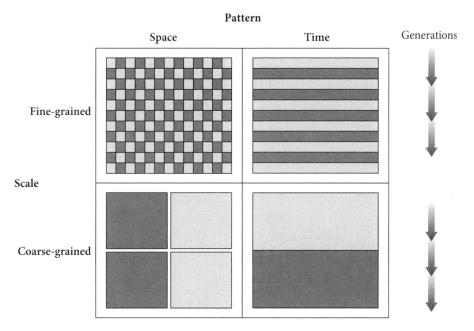

Figure 4.2 The components of environmental variation. I. Pattern and scale.

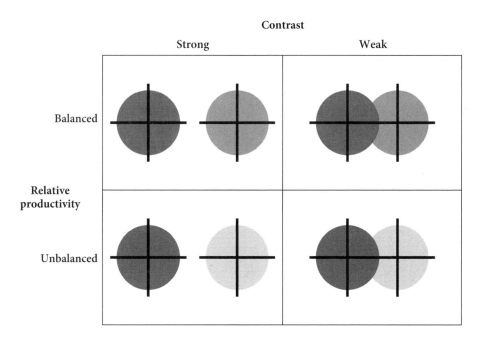

Figure 4.3 Decomposition of environmental variance. II. Contrast and productivity. Phenotypic optima that are farther away represent more highly contrasted environments (left column) than those that are closer (right column). More productive patches (dark gray) support more individuals than less productive patches (light gray).

optima, whereas those that are less highly contrasted have optima that are closer in phenotypic space.

These two descriptions of environmental variation are by no means complete. But they do capture, as we'll see, many of the essential features governing the outcome of selection in heterogeneous environments. They are also the forms of environmental variation that have been most studied in experimental systems. Nevertheless, we shouldn't lose sight of the many other ways to describe environmental variation. We could distinguish between, for example, abiotic and biotic sources of selection; resources that are depleted through population growth, like food; or those that are not, like temperature. We could also describe environmental variation itself more precisely: Is it regular and repeatable, or stochastic and unpredictable? Does it occur all at once, or does it change gradually? Some of these topics are being investigated, as we shall see, with experimental evolution. Others remain largely unexplored and represent potentially fruitful avenues for future research.

THE GENETICS OF ADAPTATION TO VARIABLE ENVIRONMENTS

Adaptation to a variable environment involves the same basic principles as adaptation to a single environment, as discussed in Chapter 2. Beneficial mutations arise and are substituted by selection, and this process continues until no further beneficial mutations are available. The process of adaptation is more complicated in a variable environment, however, because of two additional factors: gene flow and how much the conditions of growth vary in space or in time. Let's now deal with each of these factors in turn.

Gene Flow

The effects of gene flow on adaptation are easiest to see if we start by examining a situation where there is migration between subpopulations subdivided in space, but the underlying patches are otherwise identical. Consider first a situation where there's no gene flow. Selection acts locally; a beneficial mutation arising in one subpopulation is effectively stuck there. Fixation of a beneficial mutation across all subpopulations is thus prevented by population subdivision in the absence of gene flow. Provided distinct mutations are substituted in different subpopulations, genetic variation across the entire collection of subpopulations will be high and rates of adaptation slow.

However, let us imagine there is some gene flow among subpopulations. Now, a beneficial mutation arising in one subpopulation can disperse to others and a subdivided population can explore more routes to adaptation because each subpopulation, provided population sizes are large enough, is likely to produce its own beneficial mutation. In the presence of gene flow, the fittest mutation uncovered by a given subpopulation can spread to all others. Unlike a situation without gene flow, one where gene flow occurs can speed up rates of adaptation, and decrease genetic variation, because subdivided populations are better at "finding" rare beneficial mutations.

One study has directly tested the effects of variable degrees of gene flow on rates of adaptation in spatially structured populations. Kryazhimskiy, Rice, and Desai (2012) showed that increasing gene flow among subpopulations of haploid asexual populations of the yeast, *Saccharomyces cerevisae,* increased the rate of adaptation, as expected. Bell and Gonzalez (2011) found a similar result when they subjected yeast populations to severe salt stress. Evolutionary 'rescue', which occurs when a beneficial mutation arises that prevents a population from going extinct, was more likely to occur at higher rates of dispersal among subpopulations. Other experiments have contrasted rates of adaptation between the two extremes of gene flow—all or none. Perfeito et al. (2008) found that population subdivision without gene flow slowed the rate of adaptation relative to a well-mixed population in *E. coli.* Rozen et al. (2008) saw a similar effect, but only in complex environments containing multiple resources. In simpler environments containing just a single resource, migration had little effect. There are likely many more different ways a population can adapt to a complex environment, and relatively few ways of getting better in a single resource environment, so these results are in line with the idea that population subdivision can lead to a more comprehensive exploration of the underlying mutational landscape. None of these studies examined the effects of population subdivision on genetic variation.

It's worth mentioning here a close parallel between the theory just outlined and Wright's shifting balance theory of adaptation (Wright 1982). Wright felt that selection would be most effective at generating adaptation when populations were subdivided, but linked by migration. A population that eventually finds the right combination of genes will climb a local adaptive peak and be so successful that its population size proliferates, leading to dispersal into neighboring subpopulations. In this way, a successful combination of genes would spread to fixation in the population. Although this description is something of a caricature of the shifting balance process, it's important to consider if only because Wright felt it to be a very general description of how adaptation proceeded in nature. The experimental results described earlier show that, at least regarding the effects of dispersal on adaptation when populations are subdivided, Wright may not have been far wrong.

Spatial Variation

Now let's add in environmental variation underlying spatial structure. In addition to population subdivision, each population occupies a patch with a different phenotypic optimum. In Chapter 3, we considered the extreme situation where there was no gene flow among subpopulations, but conditions of growth varied for each subpopulation. Within each patch, adaptation proceeds as it does in a well-mixed population. We saw that this scenario generated strong divergent selection leading to the evolution of specialization and trade-offs in fitness across patches of the environment. How does gene flow affect the rate of adaptation and the level of genetic variation maintained across the entire population?

What happens if there is gene flow among patches? When selection is weak relative to migration, the selective advantage of a beneficial mutation depends not only on its fitness in the patch where it first arose but also its fitness across all other patches it's likely to encounter as it moves through the population (Campos et al. 2008; Débarre, Ronce, and Gandon 2013; Yeaman and Whitlock 2011). Gene flow thus allows a beneficial mutation to be "tested" across more conditions than otherwise would happen in a uniform environment. Genes with environment-specific effects are therefore expected to fix (or be eliminated) more slowly than they would be in a more homogeneous environment, slowing the rate of adaptation and prolonging coexistence of genetic variation more than would a well-mixed population (Whitlock 1996).

How the combination of migration and divergent selection affect an adaptive walk has not been studied. We can speculate that the result will depend on the pleiotropic effects of beneficial mutations. If pleiotropy is positive, so a mutation beneficial in one patch is also beneficial in another, then there should be little or no effect on the number of beneficial mutations substituted. On the other hand, if beneficial mutations tend to be antagonistically pleiotropic, or become so as they near the optimum, this may have the effect of increasing the length of an adaptive walk. Why? Because only mutations of small effect are likely to provide any advantage across multiple patches.

Tests of these predictions are almost nonexistent. Experiments with bacteria (Bailey and Kassen 2012) and phage (Cuevas, Moya, and Elena 2003) offer some limited support that rates of adaptation in spatially variable environments with gene flow are slower than in homogeneous environments. There are no direct tests of the idea that adaptive walks should be longer in spatially variable environments compared to homogeneous ones, although some preliminary evidence is suggestive: Sequencing the evolved lines from Bailey and Kassen (2012) revealed more mutations fixing in populations subjected to a regime of regular dispersal among patches composed of different carbon sources than in simple, well-mixed environments composed of just a single carbon source. Although this result is far from conclusive because sequencing was performed on whole populations rather than single genotypes, it is at least in line with the expectation that adaptive walks can be longer in spatially variable environments than in more homogeneous ones.

It's also important to remember here that the situation just described involves multidirectional gene flow, in the sense that all patches contribute at least some migrants. However, gene flow may sometimes be primarily unidirectional, for example, when a highly productive "source" population sends out migrants to other patches. If migrants are so poorly adapted to the patches where they end up that a self-supporting population cannot be maintained in the absence of dispersal, these patches are called *sinks* (Holt and Gomulkiewicz 1997). Adaptation to a sink environment can then proceed only if beneficial mutations arise first in the source population and then manage to make it to the sink population. Perron, Gonzalez, and Buckling (2007) showed that the rate of adaptation increases with the rate of migration in a source-sink model because migration effectively increases the mutation supply rate in the sink and so increases the chances that a beneficial mutation in the sink environment will be found.

Temporal Variation

In a temporally varying environment, a lineage grows first in one patch and then another; there is no escape from selection. We can effectively ignore gene flow now, because all individuals experience the same set of conditions. When the environment alternates between different states or growth conditions, we say that selection is fluctuating. These fluctuations may be periodic and cyclical, or they may be more stochastic. Temporal variation in selection also occurs when the fitness optimum changes gradually but in a consistent direction, as in Dallinger's temperature experiments discussed in Chapter 1. This sort of variation generates persistent directional selection in time.

The genetics of adaptation under fluctuating selection have not been studied in detail. We might speculate that we would see similar results to that under spatially varying selection, especially when it comes to the rate of adaptation. Fluctuating selection, after all, should make it extremely difficult for genes with environment-specific effects to fix. Compared to a temporally unvarying environment, then, fluctuating selection should slow the rate of adaptation and support higher levels of genetic variation. When the environment changes gradually, the mutations substituted are constrained to have small effects, because they can increase fitness by no more than the current distance to the optimum. Thus adaptive walks will be longer under gradual change compared to a sudden, steplike change that is the sum of all the small changes (Collins, de Meaux, and Acquisti 2007; Kopp and Hermisson 2009). This sort of model is probably a good description of environmental variation through geological time, or in response to anthropogenic changes in climate such as global warming. It might also be useful when considering how populations adapt to clinal gradients, for example, toward a point source of pollution in a river system.

What do the data say? There is evidence that rates of adaptation are often slower in fluctuating environments compared to the rates in temporally constant environments composed of the constituent, component patches (Bennett, Lenski, and Mittler 1992; Coffey et al. 2008; Vasilakis et al. 2009), although sometimes these rates are slower compared to one but not all constant environments (Cooper and Lenski 2010; Hughes, Cullum, and Bennett 2007; Jasmin and Kassen 2007b). Interestingly, data from phage experiments on rates of DNA sequence substitution in temporally varying environments consisting of alternating host-cell types and in constant environments consisting of a single host-cell type shows little effect of fluctuating selection (**Table 4.1**): 1.8 substitutions per 100 generations are fixed under fluctuating conditions versus 1.58 per 100 generations in constant ones. The difference is not significant under a one-tailed, paired t-test ($P = 0.28$). It may be that in these experiments the genes selected under fluctuating conditions are those that confer adaptation to the environmental variation itself rather than the component patches of the environment. However, the information necessary to distinguish between these alternatives is not available.

As for adaptation to gradually changing environments, the limited data that exists lends support to the idea that the mutations fixed have smaller fitness effects than would otherwise be seen under a step change in conditions. Collins and de Meaux (2009) showed that the magnitude of the first step in adaptation to decreased concentrations of phosphorous in the growth media of *Chlamydomonas reinhardtii* tended to be smaller under gradual rates of change as compared to a single abrupt change. Similarly, Perron, Gonzalez, and Buckling (2008) showed that the evolution of resistance to high concentrations of the antibiotic rifampicin in *Pseudomonas aeruginosa* occurred more readily under gradual as opposed to abrupt changes in antibiotic concentration. Similar results were observed when Bell and Gonzalez (2011) subjected yeast populations to varying rates of environmental deterioration imposed by salt stress. However, there are to my knowledge no tests of the prediction that adaptive walks are longer under gradual as opposed to abrupt change.

EVOLUTION OF THE ECOLOGICAL NICHE AND MAINTENANCE OF GENETIC VARIATION

As we've seen, environmental variation complicates the process of adaptation. It also changes the ecological characteristics of the genotypes that evolve. All else being equal, a variable environment should select for a type that has high fitness across the range of growth conditions it regularly encounters. The implication is that the range of environmental variation typically experienced

Table 4.1 Genomic changes during selection in temporally varying environments in phage

Species	Generations	Citation	Founding population	Environment	Total genomic changes	Genomic changes per 100 gens[*]	Notes
Dengue virus	200	Vasilakis et al. 2009	Isogenic	HuH-7	2.45	1.225	Generalist evolved
	200			C6/36	1.26	0.63	
	400			Alternating	2	0.5	
	200		Diverse	HuH-7	5.25	2.625	
	200			C6/36	3.75	1.875	
	400			Alternating	7	1.75	
Eastern equine encephalitis virus	1330	Weaver et al. 1999	Isogenic	BHK	7	0.526315789	Generalist evolved
				C6/36	7	0.526315789	
				Alternating	1	0.07518797	
Venezuelan equine encephalitis virus	200	Coffey et al. 2008	Isogenic strain 8131 (enzootic)	mouse	1	0.5	No adaptation to alternating environment
	200			mosquito	1	0.5	
	400			Alternating	4	1	
	200		Isogenic strain 3908 (epizootic)	mouse	1	0.5	
	200			mosquito	0	0	
	400			Alternating	3	0.75	
Vesicular stomatitis virus	100	Remold, Rambaut, and Turner 2008	Isogenic	HeLa	4.75	4.75	Generalist evolved[a]
				MDCK	5.25	5.25	
				Alternating	6.75	6.75	

[*] Mean across replicate lineages

should cause the breadth of adaptation to evolve. Highly variable environments are thus expected to favor the evolution of broadly adapted generalist types while less variable environments lead to the emergence of more narrowly adapted specialists. All else is not always equal, of course, and in some situations selection can lead to the emergence and coexistence of specialists adapted to different patches of a variable environment. Understanding the conditions underlying each of these two outcomes—the evolution of a single, broadly adapted generalist type versus the stable coexistence of more narrowly adapted specialists—has been a major focus of evolutionary ecology. Let's consider first the evolution of the breadth of adaptation or niche and then move on to consider the problem of genetic variation and coexistence.

Definition of the Niche

Before proceeding to consider how the niche evolves, we need a clear definition of what it is. The concept of the niche in ecology has a confused and sometimes tormented history. It has been taken to mean the range of conditions within which a species is regularly found and also what that species does—its role, or job—in the community. Chase and Leibold (2003) have characterized the distinction as being how the environment affects the species versus how the species affects the environment.

   Here we'll take the view that the niche is the property of a genotype (not a species, as is common in ecology) that describes the range of habitats or conditions tolerated. The niche can thus be measured by how fitness changes as a function of an environmental variable or variables. The focus on genotypes is justified, because it is the genotype that evolves in response to selection. The sum of the niches of all genotypes gives us the niche of a species. We'll take *environment* here to mean any and all factors that can affect a genotype's fitness: they might be abiotic like pH, salinity, or nutrients; or they might be biotic like competitors, predators, or parasites. This definition is general enough to account for the effects of the environment on an organism as well as the organism's effects on the environment. What matters is how these effects translate into fitness. Note that Chase and Leibold (2003) define the niche as the range of conditions where birth rates exceed or equal death rates. Since fitness is ultimately the difference between birth and death rates, the two definitions are equivalent.

The Evolution of Specialists and Generalists

The terms *generalist* and *specialist* reflect variation in the breadth of the niche. Specialists are narrowly adapted types that maintain fitness across fewer environments than more broadly adapted generalists. It's relatively easy to understand the conditions that favor the evolution of specialists and generalists. In a variable environment, the fittest type is the one that grows best, on average, across all conditions experienced. With no constraints, then, the niche breadth should evolve to match the amount of environmental variation experienced by a lineage. Specialists are thus expected to evolve in relatively constant, uniform environments while generalists should evolve in more variable ones.

It's important to be clear here about the meaning of average fitness. In a temporally varying environment where fitness changes at scales greater than a single generation, the average we need to consider is the geometric mean, not the arithmetic mean, of fitness across patches. The geometric mean is the nth root of the product of fitness values. It is a more appropriate measure of long-term fitness than the arithmetic mean, because it's more sensitive to variance in fitness through time. To see this, consider the following simple example. Imagine a lineage experiences an environment that alternates between two patches, A and B. If fitness in A is 1 and in B is 0, then as soon as the environment changes to B, the population dies. The geometric mean, which is just the square root of 1×0 (which equals 0) reflects this event; but the arithmetic mean, which is $(1 + 0)/2 = 0.5$, does not. The geometric mean is thus a more accurate reflection of the long-term mean fitness of a lineage in a temporally varying environment.

We can test the prediction that niche breadth should evolve to match the amount of environmental variation by tracking the evolution of the environmental variance of fitness, which measures how much fitness varies across a set of conditions. For a given mean fitness, a generalist has a lower environmental variance of fitness than a specialist. If the niche has evolved to match the amount of environmental variation experienced by a lineage during selection, then the environmental variance in fitness will be less following selection in a variable environment than in a constant environment.

Table 4.2 summarizes studies that have conducted such tests. The prediction seems to be largely upheld for selection in temporally varying environments, where eight out of nine studies show the expected pattern of environmental standard deviation in fitness as being greater in constant environments than in the variable one, reflecting the evolution of specialists in constant environments and more broadly adapted generalists in variable ones. Results in spatially varying environments are more variable; 10 in 16 cases match the prediction for a spatially fine-grained environment, and 3 in 5 do so for spatially coarse-grained environments.

Many other experiments have reported similar results, but haven't provided measures of environmental variance and so couldn't be included in this table. These include experiments with lines of the unicellular alga, *Chlamydomonas reinhardtii,* selected to grow as phototrophs in the light and as heterotrophs in the dark (Bell and Reboud 1997; Kassen and Bell 1998; Reboud and Bell 1997) and viral experiments examining evolution of the host range (Cooper and Scott 2001; Vasilakis et al. 2009; Weaver et al. 1999). Collectively, the results of these experiments lend support to those shown in Table 4.2: temporal

Table 4.2 Environmental standard deviation in fitness following selection in more or less variable environments

Species	Citation	Generations	Environments	Constant*	Temporal	Spatially fine-grained	Spatially coarse-grained	Notes
PHAGE								
VSV	Turner and Elena 2000	100	Host cells	2.41	0.92			HeLa, MDCK cells; alternating random
					0.81			random
	Smith-Tsurkan, Wilke, and Novella 2010		Host cells	18.21	67.92			HeLa, MDCK cells
BACTERIA								
Escherichia coli	Hughes, Cullum, and Bennett 2007	1000	pH	0.27	0.03			pH 5.3, pH 7.8
					0.02			Random variation among pH 5.3, 6.3, 7.0, 7.8
Escherichia coli	Bennett and Lenski 1993	2000	Temperature	0.13	0.01			32 C, 42 C
Escherichia coli	Cooper and Lenski 2010	2000	Carbon sources	0.04	0.06	0.01		Glucose, maltose
				0.21	0.14	0.14		Glucose, lactose
Pseudomonas fluorescens	Jasmin and Kassen 2007b	500	Carbon sources	0.32	0.22	0.47	0.07	Mannose, xylose; Spatial: unequal productivity among patches
							0.54	Spatial: productivity equalized
	Jasmin and Kassen 2007a	600	Carbon sources	0.07		0.09		Glucose, mannose; low productivity
						0.09		High productivity
				0.06		0.04		Glucose, mannitol; low productivity
						0.04		low productivity

Species	Citation	Generations	Environments	Environment variation				Notes
				Constant*	Temporal	Spatially fine-grained	Spatially coarse-grained	
				0.03		0.05		Glucose, sorbitol; low productivity
						0.03		High productivity
				0.15		0.03		Mannose, mannitol; low productivity
						0.10		High productivity
				0.08		0.02		Mannose, sorbitol; low productivity
						0.17		High productivity
				0.03		0.01		Mannitol, sorbitol; low productivity
						0.02		High productivity
	Bailey and Kassen 2012	1000	Carbon sources	0.09	0.10	0.10	0.14	Glucose, mannose, xylose

* Average environmental standard deviation for lineages selected in each of the component patches of the variable environment

variation selects for generalists while results for spatial variation tend to be more variable.

Temporal variation is therefore more likely than spatial variation to lead to the evolution of broadly adapted generalists, at least when selection acts on de novo variation arising through mutation. This result is not surprising. When the environment varies in time, there is no refuge from selection; any mutation that improves fitness in one but not the other environment will be quickly removed. Only those that increase fitness in both environments, or at least do not compromise fitness in either, stand any chance of being substituted. In spatially varying environments, on the other hand, selection on niche width is less effective because there is always a refuge from selection for types that are well adapted to some but not all conditions. The divergent selection generated by spatial variation can thus sometimes lead to the evolution of specialists, even in the face of gene flow.

Are Generalists Jacks-of-All-Trades but Masters of None?

It's commonly thought that when a generalist evolves, it must necessarily trade off high fitness in any given environment in order to maintain broad adaptation across a wide range of environments. In other words, a generalist is a "jack-of-all-trades but a master of none." Does this happen in selection experiments? Not always.

Consider the two examples in **Figure 4.4**, which illustrates the two extreme kinds of results seen. A jack-of-all-trades generalist would be one that falls well below the line connecting the fitness of the two specialists in each environment. This is exactly what we see in Figure 4.4a, which shows data redrawn from Bennett, Lenski, and Mittler (1992). They had selected a single *E. coli* clone in a temporally varying environment that alternated daily between 32ºC and 42ºC, very close to the lower and upper thermal limits for *E. coli*. On the other hand, we see the opposite result when the same ancestral strain is selected in fluctuating pH environments (Hughes, Collum, and Bennett 2007). The fitness of the two generalists (which were selected in either cyclic or randomly alternating acidic and basic conditions) falls above the line connecting the specialists: they are as good as or better than the specialist in basic media and better than it in acidic media. These "superior" generalists have therefore evolved without much compromise in fitness at all.

When are we likely to see the evolution of a jack-of-all-trades generalist, as in Figure 4.4a? When there are strong functional constraints, such that fitness effects across environments are antagonistically pleiotropic. If, on the other hand, adaptation is due mainly to the substitution of mutations that are beneficial in one environment but do not compromise fitness in other environments, then we expect that the generalists that evolve will be able to

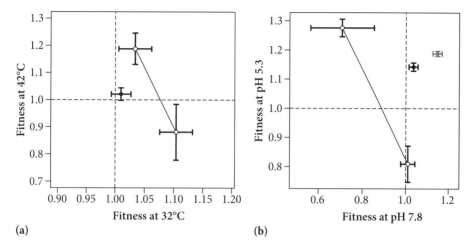

Figure 4.4 Two examples of experiments with *E. coli* that have examined whether generalists evolve to become jacks-of-all-trades. (a) A jack-of-all-trades generalist (black point). Open circles are lines selected in constant environments. (Data from Bennett, Lenski, and Mittler 1992.) (b) Universally superior generalists. Black point is a line that evolved under random alternation between pH levels, whereas the gray point is a line that evolved under daily alternation between pH levels. (Data from Hughes, Collum, and Bennett 2007). Error bars are ± 1 s.e.m.

maintain high fitness across a wide range of environments, as in Figure 4.4b. This interpretation suggests that the Fisher two-optima framework introduced in Chapter 3 might be useful as a guide for understanding when these different kinds of generalists are likely to evolve: jacks-of-all-trades generalists when the founding genotype is located between the two fitness optima, and superior generalists when the founding genotype is far from both optima. To date there have been no direct experimental tests of this idea.

Constraints on Niche Evolution

The prediction that niche breadth should evolve to match the amount of environmental variation experienced is true only if there are no constraints preventing a genotype from accessing the optimal phenotype in any environment. However, constraints can prevent this from happening. These constraints may be demographic, ecological, or genetic. We'll consider each constraint in turn.

Demographic constraints arise from variation among patches in how many individuals they produce (see Figure 4.3). Because selection is more effective in larger populations, adaptation will occur preferentially to the most productive patches. If the rate of dispersal also depends on the number of individuals a patch produces, as it does in phase 3 of Wright's shifting balance theory of adaptation and under models of so-called hard or density-independent selection (which simply means that the more individuals a patch can support in a metapopulation setting, the more it contributes to the total population), then productive patches will send out more migrants than unproductive ones, frustrating adaptation to unproductive patches (Holt and Gomulkiewicz 1997; Kawecki and Ebert 2004; Levins 1968). The combined effect of both processes should lead to the evolution of a single niche specialist on the most productive patch, even if the environment is much more variable. This was precisely the result observed by Jasmin and Kassen (2007a) when lines of *Pseudomonas fluorescens* were selected in all possible pairwise combinations of four carbon sources for 600 generations: adaptation occurred preferentially to the most productive carbon source in each pair. Moreover, selection in a temporally varying environment that alternated between a productive and unproductive resource led to the evolution of a specialist on the more productive resource, not a generalist as would normally be expected (Jasmin and Kassen 2007b). Hall and Colegrave (2007) saw similar effects of variation in productivity among patches on adaptation when they allowed *P. fluorescens* to evolve on mixtures of four carbon sources across a gradient of total resource concentrations. Adaptation to the most productive resource occurred at the highest total concentrations but, notably, the community supported a mixture of genotypes at intermediate concentrations and a single, broadly adapted generalist at very low concentrations. Presumably these different outcomes reflect how the relative productivity of the different component resources changes as their concentration changes, although this was not something directly measured in the experiment.

Ecological constraints are caused by the interaction of a focal lineage with other species or genotypes. Such constraints come in many flavors. The presence of a predator, for example, can reduce prey population densities enough to drastically slow or even prevent the prey invading novel niche space (Buckling and Rainey 2002b; Meyer and Kassen 2007). Interspecific competitors can also prevent occupation of a novel niche by a focal lineage, as evidenced by the many experiments in fruit flies (Bolnick 2001), sticklebacks (Bolnick et al. 2010), and microbes (Brockhurst, Colegrave, and Hodgson 2007a; Fukami et al. 2007; Zhang, Ellis, and Godfray 2012) showing that when competitors are removed the range of the focal lineage expands. It has also been suggested that adaptation to the presence of a competitor might come at a cost to adaptation to the physical conditions of the environment. Collins (2011), for example, has shown that the presence of a competitor slowed adaptation to changing CO_2 conditions in the unicellular green alga *Chlamydomonas reinhardtii,* consistent with this idea.

Genetic constraints come about because the appropriate genetic variants that improve fitness across multiple or novel environments are unavailable to selection. This can happen because, in mutation-limited populations, the right allele or combination of alleles may not be accessible to the current genotype; or, in populations containing standing genetic variation, negative genetic correlations in fitness across environments prevent a response to selection to multiple environments. When propagating viruses on different host cells, it is not uncommon for investigators to see adaptation to single hosts in isolation but not to a regime of alternating host-cell types (Coffey et al. 2008; Vasilakis et al. 2009). Presumably this happens because a single mutation cannot improve fitness on both hosts, and so the population must wait for multiple mutations to be substituted before adaptation to a variable environment occurs. Notably, the treatment of HIV/AIDS has benefited immensely from this kind of constraint. Highly active antiretroviral therapy (or HAART), the standard treatment regime for HIV treatment in the developed world, is a cocktail of drugs with very different functional targets. Viral resistance to each drug occurs fairly readily, but for the virus to become resistant to the therapy, it must be resistant to all drugs in the cocktail simultaneously. If resistance is already a fairly rare event occurring with probability, μ, the probability that a single virus is doubly resistant is μ^2, triply resistant μ^3, and so on. Thus combination therapies like HAART are effective at reducing viral population sizes while at the same time effectively preventing the evolution of resistance (Chow et al. 1993). While it is well-recognized that genetic correlations can often constrain responses to selection in multicellular organisms (Walsh and Blows 2009) it's worth noting that genetic correlations can limit the response to selection from standing genetic variance in microbes as well (Bell 1997b; Scheiner and Yampolsky 1998).

Genetic constraints can also arise because the right gene or genes are simply not present or because different substitutions functionally interfere with each other in such a way that fitness improvements in one environment are associated with fitness losses in another. *Pseudomonas fluorescens* strain SBW25 lacks the gene *xylB,* a central component of the xylose operon and, as a result, grows just barely on xylose—a trait that's diagnostic for this strain. Interestingly, selection in a xylose-containing medium rapidly restores the ability to grow on xylose (Bailey and Kassen 2012; Jasmin and Kassen 2007b; Melnyk and Kassen 2011), although the mechanism that allows this to happen has not been worked out. Similar results have been seen for other substrates in other species (Blount, Borland, and Lenski 2008; Duffy, Turner, and Burch 2006; MacLean and Bell 2002; Mortlock 1983), suggesting that what at first appears to be a fairly hard constraint is not so hard that selection cannot break it, eventually.

Functional interference between mutations at different genes, or even within the same gene, can also prevent niche expansion. Duplication of the *lac* gene in *E. coli* allows adaptation to lactulose, while mutations in *mgl* allow adaptation to methyl-galactosidase (Stoebel, Dean, and Dykhuizen 2008).

Both mutations incur fitness costs on the alternative substrate. As a result, chemostat cultures containing both resources can lead to the emergence and coexistence of lactulose and methyl-galactosidase specialists for up to hundreds of generations and apparently prevent the emergence of generalists (Dykhuizen and Dean 2004; Zhong et al. 2004). Other examples of antagonistic pleiotropy underlying the coexistence of specialists include the radiation of *P. fluorescens* in static microcosms (Rainey and Travisano 1998) and the emergence of acetate and glucose specialists in *E. coli* (Friesen et al. 2004; Spencer et al. 2007).

Evolution of the Multidimensional Niche

We saw previously in Chapter 3 that divergent selection invariably leads to specialization and fitness trade-offs across environments. Sometimes these trade-offs are associated with a cost of adaptation—lower fitness than the ancestor in an alternative environment, and sometimes not. This description of the niche must certainly be too narrow, since we are considering the impact of adaptation on only one or at most two environments. What happens, then, if we expand our view to include a wider range of environments?

Adaptation to one environment often does come at the cost of a narrower niche. Maughan et al. (2006) showed that when lines of *Bacillus subtilis* were selected for 6000 generations in a rich medium containing all the necessary macromolecules required for growth and replication, they often lost the ability to synthesize key vitamins and resources due to loss-of-function mutations in unused biosynthetic pathways. Similarly, *E. coli* selected for 20,000 generations at 37°C in a medium that contained only glucose as the sole energy source lost the ability to grow on a wide range of alternative carbon sources (Cooper and Lenski 2000), and its thermal tolerance was narrower than that of the ancestor it was derived from (Cooper, Bennett, and Lenski 2001). Other experiments show comparable results: prolonged selection in a single environment leads to a narrowing of the niche relative to the ancestral niche when a wide range of environments are considered (Alto and Turner 2010; Bennett and Lenski 2007; Collins and Bell 2004; Cooper 2002; MacLean, Bell, and Rainey 2004; Travisano et al. 1995).

It has been occasionally observed that adaptation to one environment leads to niche expansion rather than niche narrowing. MacLean and Bell (2002), for example, showed that *P. fluorescens* selected for 1000 generations for specialization on distinct carbon substrates lost the ability to grow on 2 to 3 other substrates that its ancestor could grow on. This finding is consistent with the evolution of specialization, but surprisingly, the evolved lines also gained the ability to grow on about 30 novel substrates the ancestor could not use. Cooper (2002) observed a similar result in the long-term selection lines of *E. coli*. This kind of niche expansion is probably due to adaptation to the general conditions of growth in the laboratory. Possibly, for example, some mutations involved in adaptation to the laboratory have the pleiotropic effect of generally increasing fitness. For example, whole genome sequencing of our own evolved lines of *P. fluorescens* often reveals many mutations—from single base-pair changes up to large-scale deletions of tens of thousands of base pairs—that lead to a loss of motility. This result is not surprising, because our strain is normally motile but our experiments are performed in shaken liquid cultures where motility is probably useless. Losing these costly motility-associated functions likely leads to increases in fitness in shaken cultures, regardless of the range of resources available. Other mutations that could have similarly broad effects on fitness across different environments include those involved in global regulation of gene expression in stressful environments (Crozat et al. 2005; Ferenci and Spira 2007). To the extent that such changes indicate patterns of adaptation to novel environments more generally, the early stages of adaptation might often be associated with a modest broadening, rather than a narrowing, of the niche. This hypothesis deserves closer attention.

Tolerance Curves

The biochemical and physiological conditions of life set certain limits on the range of conditions an organism can tolerate. Beyond these limits, proteins cannot fold properly, enzymes bind less effectively to substrates, and the basic metabolic processes of life start to break down. The response of fitness across a gradient of a particular environmental variable, like salinity, temperature, or pH, is called a tolerance curve. Tolerance curves are often described as being humped, and fitness as being maximal at some intermediate value of the environmental variable and falling off toward the extremes (see, e.g., Knies et al. 2006). However, other functions may be more appropriate depending on the particular environmental variable being considered.

Under this view, the limits to tolerance—the niche boundaries, as it were—are defined by functional trade-offs that generate antagonistic pleiotropy. Any improvement in fitness beyond the normal tolerance limits should come at a cost to tolerance in other parts of the tolerance curve. Mongold, Bennett, and Lenski (1996) found a result consistent with this prediction: selection for cold tolerance in *E. coli* at a temperature just one degree above its lower thermal tolerance limit resulted in a decrease in the upper thermal tolerance limit by 1–2°C. Other studies have had less success in breaking the thermal tolerance limits in *E. coli* (Bennett and Lenski 1993; Knies et al. 2006). The main effect of selection at or near these boundaries was to shift the optimum temperature toward the temperature of selection.

The biochemical and physiological details that underlie these changes in the tolerance curve have not been studied in detail. This seems like an interest-

---

BOX 4.1

Quantifying the Niche

Dictionaries define the word *niche* as a shallow recess in a wall that often holds a statue or other decorative object—a sort of small, built-in shelf space. Ecologists co-opted the term in the early twentieth century using it to mean either an organism's role or function in the community, or the range of habitats or conditions that an organism can tolerate. The latter interpretation is emphasized here, because it provides a more direct link to measurable environmental variables.

In thinking about how to quantify the niche, it's useful to recognize that each environmental variable represents a distinct axis that, collectively, describes the conditions within which a genotype or population can (or does) exist. This approach comes closest to the n-dimensional volume concept of the niche developed by Hutchinson (1957). We can further describe the shape and width of the niche along a single axis of environmental variation. This is the tolerance curve approach discussed in the main text.

Reaction Norms

A reaction norm is a function that relates how phenotype changes as a function of an environmental variable for a given genotype. The phenotype can be any kind of trait, including fitness. If there are just two environments, this function will necessarily be linear, as illustrated in panels (a) and (b) of the following figure. The black and gray lines represent the reaction norms for two genotypes whose fitness is measured in two environments, E1 and E2. The black line represents a more specialized genotype, because the slope of its reaction norm is steeper than that of the "gray" genotype. In panel (a), the specialist is fitter than the generalist in both environments; diversity cannot be maintained by divergent selection. In panel (b), the reaction norms cross such that the gray genotype is fittest in E1 and the black genotype is fittest in E2. This is a necessary condition for diversity to be maintained.

(continued)

---

ing avenue for future research, because it could tell us something about the consequences of environmental change on the physiological functioning of organisms. More generally, this work emphasizes that we still do not understand how the tolerance curve evolves: Is it something that's evolutionarily labile, or is it a more rigid, robust property of a genotype? The answer to this question, and more precise details of why and how the tolerance curve changes when it does evolve, represent promising avenues for future research.

It's common to use analysis of variance to decompose the phenotypic variation in reaction norms into components due to environmental variation, genetic
variation, and their interaction. Adding more environments can make analysis
more complicated, but the principles are the same. The genotype-by-environment interaction can be further decomposed into components due to differences
in the extent of genetic variation expressed in each environment (panel a) and
to changes in the rank order of fitness among genotypes across environments
(panel b; note the crossing reaction norms; Robertson 1959; Venail et al. 2008).
An introduction to the analysis of reaction norms and the decomposition of
genotype-by-environment interactions is provided in Lynch and Walsh (1997),
and their role in the evolution of phenotypic plasticity and the maintenance of
genetic variation in fitness is discussed in Via and Lande (1985) and Via et al.
(1995).

Multidimensional Niches

The simplest way to describe a multidimensional niche is as the mean and variance of fitness across conditions of growth. For a given mean fitness, specialists
display a higher variance in fitness across environments than generalists do. This
result has the advantage of being relatively simple to understand and compute.
The disadvantage is that the ordering of the environments is often arbitrary and
so limits the range of statistical tools available to analysis of variance. For this
reason, agronomists often use the mean fitness of all genotypes in a given environment as a measure of that environment's quality, in a method called stability
analysis (Finlay and Wilkinson 1963). Regressing the fitness of a particular genotype against the mean of all genotypes thus provides a relative measure of niche
breadth: specialists are more responsive (have a steeper slope) than generalists.
This situation is illustrated in panel (c), where the dashed line represents the 1:1
curve, that is, the slope of an "average" genotype regressed against the mean fitness of all genotypes in each environment. The black line indicates a less stable
type that is a specialist in "good" environments (those that have higher average
fitness). The gray line is a more stable generalist that does better than average in
poor environments but does less well than the average in good environments.

(continued)

ENVIRONMENTAL HETEROGENEITY SUPPORTS GENETIC VARIANCE IN FITNESS

The primary effect of environmental variation is to change the direction of
selection, either in space or in time. The main effect of environmental variation will therefore be to slow the rate at which mutations with environment-
specific effects either sweep to fixation if they are beneficial or are eliminated

BOX 4.1 (continued)

Tolerance Curves

Tolerance curves are special kinds of reaction norms where there is a specific function that describes how fitness is expected to change at different values of the environment. Tolerance curves for temperature, for example, are often humped, and fitness is highest at some intermediate value of the environmental variable, as illustrated in panel (d). The figure also shows a specialist, represented by the black curve, and a more broadly tolerant generalist, represented by the gray curve.

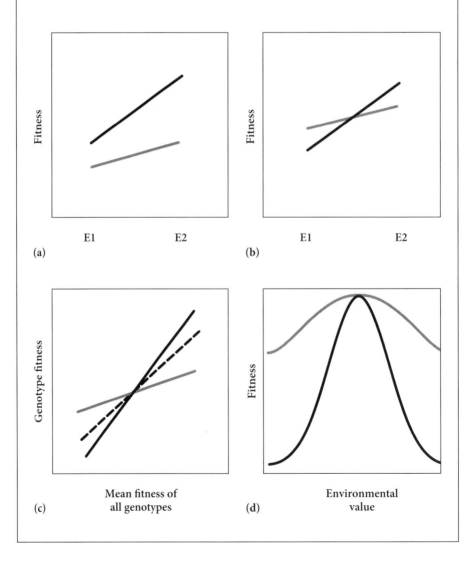

if they are deleterious. Although this genetic variation is ultimately transient, it will tend to cause genetic variance in fitness to be higher in heterogeneous environments than in more homogeneous ones. Moreover, because spatially varying environments afford refuges for specialized types but a temporally varying environment does not, diversity should be maintained more readily in spatially varying environments than in temporally varying ones, provided divergent selection is strong relative to the rate of gene flow among patches (Felsenstein 1976; Hedrick 1986).

Environmental heterogeneity can support diversity indefinitely as well. This happens either because different types are fittest in different patches of the environment or, for the special case of sexual diploids, heterozygotes enjoy a fitness advantage that homozygotes do not across all patches. The conditions that lead to the stable maintenance of diversity through divergent selection can seem rather restrictive: gene flow must be weak relative to selection and, most important, population regulation must occur locally, at the level of the patch, rather than globally, at the level of the total population (Levene 1953; Nagylaki and Lou 2006). This can seem like a rather idiosyncratic detail, but it is crucial for the stable maintenance of diversity because it generates negative frequency-dependent selection—rare types have higher fitness than common types. The reason is that when population regulation occurs locally, each patch contributes some fraction of its population to the next generation. Highly specialized genotypes restricted to just one patch are thus assured of being represented in the total population. Global population regulation, on the other hand, leads to replacement of the type that is fittest in the patch that contributes most to the population (Dempster 1955), a situation reminiscent of phase 3 of Wright's shifting balance theory.

Genetic Variance in Heterogeneous and Homogeneous Environments

A substantial number of experiments have tested the prediction that genetic variance in fitness should evolve to be higher in heterogeneous than in homogeneous environments. **Table 4.3** summarizes the studies that have tested this prediction when all genetic variation is introduced to an evolving population via mutation. In 12 of 19 cases, selection in the heterogeneous environment supported more diversity than selection in a homogeneous environment, but this is not formally significant in a binomial test ($P = .10$). We can eliminate one of these negative studies on the grounds that no trade-off evolved even under purely divergent selection, so we would not expect to see diversity maintained under heterogeneous conditions anyway (Jasmin and Kassen 2007a), thus giving a ratio of 12:18. It's notable that all but one of the remaining

Table 4.3 Genetic variance in fitness in heterogeneous environments

Species	Generations	Citation	Form of environmental variation	Scale of variation	Environments	Genetic variance greater in heterogenous environments?	Diversity measure	Notes
BACTERIA								
Pseudomonas fluorescens	900	Barrett, MacLean, and Bell 2005	Spatial	Fine	Mixture of 8 unspecified sugars	yes	GxE	
	~50	Rainey and Travisano 1998	Spatial	–	Static vs shaken microcosms	yes	Colony morphology, invasion from rare	
	600	Jasmin and Kassen 2007a	Spatial	Fine	Pairwise combinations of mannose, glucose, mannitol, sorbitol	yes in 5/6	GxE	
	~400	Jasmin and Kassen 2007b	Spatial	Fine	Mannose, xylose	no	GxE	Dynamics of adaptation are largely driven by adaptation to xylose
			Spatial	Coarse	Productivity equalized	no		
			Spatial	Coarse	Productivity not equalized	no		
			Temporal	Coarse		yes		
	1000	Bailey and Kassen 2012	Spatial	Fine	Glucose, mannose, xylose	no	GxE	Dynamics of adaptation driven by adaptation to xylose; note that spatial structure does influence rate of adaptation, and that adaptation occurs earlier in the mixture relative to the spatially separated treatment
			Spatial	Coarse		no		

Species	Generations	Citation	Form of environmental variation	Scale of variation	Environments	Genetic variance greater in heterogenous environments?	Diversity measure	Notes
Escherichia coli	550	Dykhuizen and Dean 2004	Spatial	Fine	Lactulose, methyl-galactosidase	yes	Polymorphism with linked phenotypic marker	
	900	Habets et al. 2006	Spatial	Coarse	Agar plate without dispersal	yes	Metabolic profile, invasion from rare	
					Agar plate with dispersal	no		
Ralstonia sp.	1000	Korona et al. 1994	Spatial	Coarse	Agar plate without dispersal	yes	Colony morphology	Identified as Comamonas in original paper
EUKARYOTES								
Chlamydomonas reinhardtii	~ 500	Bell and Reboud 1997	Spatial	Coarse	Light, dark	yes	Genetic variance in fitness within each environment	

negative cases are from one lab (mine) and have involved adaptation to xylose, a substrate that our founding strain cannot readily metabolize but can evolve quickly to do so. The dynamics of variation in these experiments are thus dominated by adaptation to xylose.

Many other experiments have examined the effect of environmental variation on the fate of diversity from genetically diverse founding populations or collections of ecologically similar species, or they have focused on the range of conditions required for coexistence among two or three well-defined types (reviewed in Kassen 2002; see also Kerr et al. 2002). All studies find broadly similar results: diversity tends to be supported more readily in heterogeneous than in homogeneous environments. It's notable that temporal variation has been observed to support diversity at levels significantly higher than constant environments in some experiments (see also Venail et al. 2011). It may be that most of this variation will eventually be eliminated, albeit very slowly. Alternatively it is possible that temporally varying environments can stably support diversity through a phenomenon known as the storage effect (Chesson 2000). The idea here is that a genotype can prevent elimination under bad conditions by waiting things out until conditions improve again, perhaps by entering a resting stage. To date the impact of storage effects on diversity has not been studied in microbial evolution experiments, although it remains an attractive explanation for the maintenance of diversity in many systems (Hairston Jr., Ellner, and Kearns 1996).

Mechanisms That Maintain Variation

Environmental variation can support genetic diversity in two ways. It can slow the rate at which genetic variation is eliminated by selection, or it can maintain diversity through negative frequency-dependent selection. In the former case diversity is ultimately transient, in the sense that variation cannot be stably maintained but must be continually reintroduced either by mutation or migration. Negative frequency-dependent selection, on the other hand, can act to maintain diversity stably because genotypes have higher fitness when rare than when common and so are protected from loss.

Few studies have sought to document the contribution of transient diversity to the maintenance of genetic variation. One exception was Elena and Lenski (1997), who nevertheless showed that the modest levels of genetic variance maintained in the otherwise spatially homogeneous conditions of a test tube containing a single carbon source were maintained by negative frequency-dependent selection, ultimately shown to be due to cross-feeding (Rozen and Lenski 2000). Bell (1997b) showed that the manner of population regulation, whether it occurred at the level of the patch or at the level of the total population, had little effect on the quantity of diversity maintained over 50 generations of selection in a spatially heterogeneous environment in initially genetically diverse populations of *Chlamydomonas reinhardtii*. This result suggests that the purportedly stabilizing effect of local density regulation made no contribution to the maintenance of diversity.

There are a growing number of examples of diversity being maintained by negative frequency-dependent selection. The canonical case is the radiation of *P. fluorescens* into distinct niche specialists that occupy different regions of static glass microcosms (described in Chapter 1). Other examples involve some sort of resource specialization that evolved in multiple-resource environments (Dykhuizen and Dean 2004; Friesen et al. 2004; MacLean, Dickson, and Bell 2005), through the creation of novel resources produced as secondary metabolites that leak into the medium and become available for growth—a phenomenon known as cross-feeding (Rosenzweig et al. 1994; Treves, Manning, and Adams 1998), or through scavenging of cellular detritus (Rozen et al. 2009). What these examples have in common is that strong trade-offs in fitness, often underlain by antagonistic pleiotropy, evolve through specialization on different resources.

CONCLUSIONS

Stripe rust became a problem in Australia because, like pathogens everywhere, it evolved. What makes the story especially compelling from an evolutionary perspective was that the founding genotype was almost certainly isogenic, or pretty close to it, and all genetic variants since appear to have emerged in situ. Pathotype variation in Australian stripe rust must therefore have been driven primarily by the combination of mutation and selection in the "test tube" of Australian wheat fields. Stripe rust evolution in Australia thus constitutes a sort of natural experiment that we can use to help validate our models for adaptation and genetic variation.

The leading result is that the dynamics of genetic variation in stripe rust across Australia certainly do not resemble the periodic selection associated with the simplest models of adaptation. Substantially more genetic variation persists for longer periods of time in Australian stripe rust than we would expect. The results of experimental studies in the laboratory tell us that incorporating more realistic features of natural populations like spatial structure and environmental variation can increase the amount of genetic variation supported during adaptation. It's no great leap, then, to suggest that perhaps environmental variation plays a role in governing the dynamics of genetic variation in stripe rust as well.

Adaptation in variable environments is a far more complicated process than it is in homogeneous, well-mixed conditions. Laboratory experiments have taken us some way toward understanding adaptation to a variable environment, however, because we can impose a particular kind of environmental variation and examine its effects. This has been the focus of Chapter 4. What have we learned?

1. We can decompose environmental variation in any number of ways; but most useful, from the perspective of adaptation, is into components of pattern (space versus time) and scale (fine versus coarse). It's also useful to remember that environments differ in their affect on the strength of selection (contrast) and demographics of subpopulations (productivity).

2. The main effect of spatial structure and environmental variation is to slow the rate of adaptation. Genetic models of adaptation to variable environments, particularly those focusing on how environmental variation affects the length of adaptive walks, remain understudied, both theoretically and experimentally.

3. Selection also influences how the ecological characteristics of genotypes, especially their niche breadth, evolve. In general, the breadth of the ecological niche should evolve to match the amount of environmental variation. Specialists therefore tend to evolve in constant environments, generalists in more variable ones. Temporal variation is especially effective at selecting for generalization; spatial variation is less so.

4. The evolution of generalization is not necessarily constrained by a trade-off between jack-of-all-trades and master of none; superior generalists that do not show such a trade-off can evolve. Which of these two scenarios governs the evolution of generalization likely depends on where in the fitness landscape a founding population is located relative to the fitness optima: if between them, then most mutations will have antagonistic effects and the result will be a jack-of-all-trades; if far from both optima, superior generalists that show fitness improvements on the component patches are likely to evolve because beneficial mutations display positive pleiotropy.

5. Demography, ecology, and genetics can all act as constraints on niche evolution, leading to situations where the breadth of adaptation evolves to be narrower than would otherwise be expected based on the amount of environmental variation a lineage has experienced. Demography in particular can be an extremely important factor in governing niche breadth, for selection often is biased toward promoting adaptation to the most productive patch.

6. As a general rule of thumb, selection in a constant environment leads to a narrowing of the multidimensional niche—that is, those other environmental variables not directly under selection. Occasionally selection for specialization leads to an increase in fitness in some aspects of the multidimensional niche, although these are usually associated with common adaptations to growth in a novel environment. In the laboratory, these mutations often involve a form of

domestication to life in the test tube. Similar processes of domestication may underlie adaptation to novel environments in more natural systems as well.

7. The evolution of tolerance curves—how fitness changes along a gradient of a single niche dimension—remains understudied. How rigid are the niche boundaries of a genotype and, when these boundaries are crossed, how is fitness at other points along the tolerance curve impacted?

8. Genetic variation is often, though not always, higher in heterogeneous than in homogeneous environments. When diversity is higher in spatially heterogeneous environments, it may be transient or stably maintained by negative frequency-dependent selection. Surprisingly, temporal variation can also support higher levels of genetic variation, although much of this diversity is likely to be transient.

These results are beginning to provide a clearer picture of how environmental variation impacts adaptation. In terms of how diversity evolves two important points emerge from this chapter.

The first is that we expect to see abundant genetic variation in a population during adaptation to a variable environment. This is true especially in spatially varying environments, but also to some extent in temporally varying ones. The implication is clear: we should not be surprised to see substantial amounts of genetic variation being maintained in natural populations. The real question to be answered is whether that variation is transient or stable. There are examples of both from microbial selection experiments, but it is not at all clear which is more common. It would be worth paying closer attention to the dynamics of genetic variation in these experiments. How often is the genetic variation supported by environmental heterogeneity used to adapt to future changes in the environment? Will transient diversity evolve to become stably maintained? These questions are fundamental to our understanding of the nature of diversity and how it evolves. Integrating whole genome sequencing into the experiments will be a valuable step forward, but we can't forget that the most important variable to keep track of is the genetic variance of fitness and how this changes across environments. Developing reliable, automated, high-throughput assays to estimate genetic variance in fitness within populations would be a step forward in this regard.

It would also be helpful to get a better idea regarding what sorts of environmental change natural populations are experiencing. How variable is nature in space relative to time, for example? One way to approach this question is through reconstruction ecology—using resting spores from lake sediments or permafrost, for example, to get a picture of how fitness of different genotypes varies through historical time. Another way is to perform reciprocal transplant experiments in natural systems. My research group, for example, has begun studying soil isolates of *Pseudomonas* taken at different locations in a Canadian

mixed deciduous forest throughout the growing season and then doing reciprocal transplants using soil-water samples in the laboratory as their environments. The method is crude; but, as Koskella (2013) and Belotte et al. (2003)
have shown, it is effective at giving some glimpse into how natural populations
of microbes, at least, are structured and perceive their environment.

The second important point bears on the evolution of the niche. Laboratory experiments suggest that niche evolution can be quite labile, evolving to match the amount of environmental variation experienced. This result
stands in contrast to comparative work that sees the niche as being conserved
through evolutionary time (Crisp et al. 2009). The difference between these
two views might be explained by the fact that constraints due to demography,
ecology, or genetics can prevent niche evolution or bias it in a particular direction, thus leading to apparent niche conservatism. The relative importance of
environmental variation versus constraints in contributing to apparent niche
conservatism remains an open issue.

There remain other gaps that also deserve attention. Different kinds of
environmental variation are expected to lead to different kinds of adaptation,
for example, but this subject has been little studied. Fine-grained variation
favors the evolution of a type that is versatile, in the sense that the phenotype responds reversibly (and appropriately) to prevailing conditions. Coarse-
grained variation leads to the evolution of a plastic type whose phenotype
responds early in development to prevailing conditions (since these are predictive of what is likely to be experienced throughout life) and remains fixed
thereafter. An environment that varies temporally in an unpredictable way
should lead to a bet-hedging strategy, where a single genotype produces a
range of phenotypes each generation but without responding in a specific way
to the prevailing conditions. Some attempts have been made to test these ideas
(Engelmann and Schlichting 2005; Kassen and Bell 1998; Suiter, Bänziger,
and Dean 2003; Beaumont et al. 2009), but little progress has been made. It
may be that such sophisticated adaptations require concerted, long-term selection to evolve. Whole genome sequencing of an evolved bet-hedging genotype
of *P. fluorescens,* for example, revealed that nine mutations were involved, and
the last mutation was the crucial one conferring the bet-hedging phenotype
(Beaumont et al. 2009). Whether this result is typical of other types of adaptations to variable environments remains to be seen.

The long-term stability of polymorphisms supported by environmental
heterogeneity is also something of a mystery. Little is known about this area,
because most experiments involving selection in heterogeneous environments
have not extended much past a couple of thousand generations. Can these
sorts of polymorphisms be destabilized by mutation, or does continued adaptation tend to promote further specialization within a niche? Results from
two short-term experiments suggest the latter. The polymorphism for resource

use in *E. coli* chemostats growing on lactulose and methyl-galactosidase has been maintained for up to 700 generations (Zhong et al. 2009), and Meyer et al. (2011) showed that the polymorphism between biofilm-forming and broth-colonizing morphotypes of *P. fluorescens* in static microcosms is maintained and accompanied by continued adaptation of each niche specialist to its respective niche. Whether these results hold up over the evolutionary long term remains to be seen.

CHAPTER 5

Genomics of Adaptation

This is Eva Markvoort: actress, poet, and artist. Her blog, 65_ redroses, is about life with cystic fibrosis. Here's a sample:

> I was in the hospital a while ago very sick, high fever, on oxygen...
> still unable to catch my breath even just lying in the hospital bed. My
> boyfriend Greg came to visit and he brought a book to read to me
> to get my mind off being so sick. He brought *The Power of Now,* an
> inspirational book about living for the moment. I'm sure this book has
> helped many people but well...I'll give you the gist of the first chapter.
> "There are no problems in your life...all you have to do is take ONE
> DEEP BREATH." I lost it. I laughed so hard tears poured out of my

Figure 5.1 Artist Eva Markvoort. Photo by Cyrus McEachern.
Courtesy of 65RedRoses.com #4Eva.

eyes…take a deep breath and your problems would be gone? If I could take a deep breath I wouldn't HAVE any problems.

(http://65redroses.livejournal.com/21911.html, from a speech posted May. 3rd, 2007)

Eva's writing is sometimes painful, occasionally humorous, and always honest. Her story is so compelling, touching, and important that it has been made into an award-winning film. In 2010 she died while waiting for a lung transplant. She was 25.

People who have it describe cystic fibrosis (CF) as like drowning on the inside. Their lungs become thick with mucus that is impossible to clear, making even breathing a challenge. Lungs are also nutrient-rich, warm, and moist environments, ideal for infection. As a result, the leading cause of death in adults with CF stems from complications associated with infection.

It was not always this way. Seventy years ago, most individuals with CF died as infants or young kids, often from problems associated with the digestive tract. A combination of improved detection and better management has both prolonged the life span of people who have CF and improved their overall quality of life. Today, more adults are living with CF than there were even a few decades ago. However, the longer these individuals live, the more likely they are to contract an infection.

Infections in those who have CF can be managed, to a point. Eventually most adults with CF end up contracting chronic infections that are impossible to clear. In the majority of cases, these chronic infections are due to a single species, *Pseudomonas aeruginosa*. Once an infection happens, the only hope for survival is a lung transplant. Eva battled with *Pseudomonas* a number of times, according to Cyrus McEachern, a doctor and photographer (he took the chapter-opening photo). Eva received one lung transplant, but died while waiting for a second.

How and why do most people living with CF become chronically infected with *P. aeruginosa*—often to the exclusion of other well-known opportunistic pathogens like its close relative *Burkholderia cepacia, Staphylococcus aureus,* or the mold *Aspergillus fumigatus?* The reason is not well understood. Clearly, a lot is going on during the course of infection. Over time, a regular succession of species infect the person with CF. *Pseudomonas* becomes dominant only once the individual reaches adulthood (Harrison 2007), and there is good evidence that *Pseudomonas* populations within and among hosts undergo substantial adaptive evolution over the course of an infection (Smith et al. 2006; Yang et al. 2011). Whole genome sequencing has identified a common set of genetic changes associated with chronic infection of lungs affected by CF, including mutations conferring antibiotic resistance, reduced virulence (probably an adaptation to avoid immune detection by the host), and biofilm formation.

Most readers probably won't be surprised that genetic adaptation occurs during the course of a chronic infection. Pathogen population sizes in the lung can be quite large, making it easy for selection to do its job. Clinical microbiologists, on the other hand, did not expect adaptive evolution to play such an important role during an infection, because they traditionally have thought of different species of microbes as being fairly immutable, having well-defined characteristics and properties that, for example, allow us to identify them as distinct species. *P. aeruginosa,* for example, is "known" to be a good pathogen, in part because it contains many virulence factors, traits that cause damage to host tissues and aid in establishing acute infections. Isolates from chronically infected CF lungs showed genetic signatures of reduced virulence, however. Chronic infection therefore seems to involve genetic changes that go against type, so to speak. This finding demands that we rethink the importance of adaptive evolution during chronic infections.

Part of this rethinking, for *P. aeruginosa* at least, will involve a closer examination of the selective environment in acute and chronic infections (see, e.g., Wong and Kassen 2011). The two situations are quite different from the perspective of the pathogen. Acute infections are about colonization and rapid growth; chronic infections also require an initial colonization event, of course, but also long-term persistence. If acute and chronic infections represent substantially different selective pressures, then different mutations are likely to be selected under each condition. The implication is that what might work in treating acute *Pseudomonas* infections of, say, a wound or urinary tract, is unlikely to work in the CF lung. In other words, we cannot rely on what we know about *P. aeruginosa* as an acute pathogen to inform treatment of *P. aeruginosa* as a chronic pathogen of CF lungs.

Much basic natural history has to be done before we can move much further beyond speculation. A first step will be to identify the genetic changes involved in adaptation to different kinds of infections. This is particularly important in the case of chronic infections of the CF lung, because it will allow us to develop more effective interventions for treatment. Doing so requires that we know more about what genes are the targets of selection and how repeatable that process is across different patients. We know comparatively little about the genomics of adaptation in general, let alone in chronic lung infections, because until recently we haven't had the ability to examine closely and comprehensively the suite of genetic changes that evolve during adaptation. This situation is changing now, with the introduction over the past few years of cost-effective whole genome sequencing technologies. This chapter reviews those experiments, primarily in phage, bacteria, and yeast, that have combined laboratory studies of adaptation with whole genome sequencing in an effort to identify general patterns and principles of adaptation at the genomic level.

The search for the genetic changes responsible for adaptive evolution has been the "holy grail" of adaptation research. With the introduction of cost-effective next-generation sequencing (NGS) technology over the past few years, the grail is finally within reach. Combining NGS with experimental evolution of microbial populations is particularly promising in this regard because it provides a glimpse into the natural history of evolving genomes under at least one fairly well-defined set of parameters: large population sizes, asexual reproduction, and (usually) haploid genomes. It's possible simply to compare the genomes of ancestral and evolved strains and identify the genetic differences between the two. The result is a fairly comprehensive list of the mutations that have been substituted over the course of evolution, a sort of genomic natural history.

Making sense of this data has, however, proven far more difficult. Why do some genes have mutations and others do not? Are all the mutations adaptive, or might some be neutral or even mildly deleterious, having hitchhiked to high frequency alongside a beneficial mutation? How important are single-nucleotide changes versus other kinds of mutation like gene duplications, insertions, or deletions in generating adaption? Answering these and other more directed, hypothesis-driven questions about the genomics of adaptation is the goal for this chapter.

We'll begin by looking at some basic numbers and examining some broad patterns. How many mutations are typically substituted in a single genome, and at what rate? Does the rate of substitution change over time? How are these mutations distributed throughout the genome? Are they typically adaptive or nonadaptive? We then consider the variety of genetic routes to adaptation in independently evolving lines, because this speaks directly to how repeatable or predictable adaptive evolution might be. In particular, we consider two questions: How common is parallel evolution, the repeated evolution of the same genetic changes in independently evolved lines? And, are any general rules available to guide us in thinking about what sorts of genes are likely to be involved in adaptation to novel environments?

This chapter is based on data provided in **Table 5.1** that comes mainly from two sources. One is a review of patterns of genomic evolution in laboratory selection experiments of the commonly used microvirid phage ΦX174 (Wichman and Brown 2010). The other is a recent review that considered genome evolution in selection experiments with bacteria and yeast (Dettman et al. 2012). This chapter updates and extends this review, and the author thanks his coauthors on that paper for their insight and time. Wherever possible, other data is included; the bulk of this comes from RNA phages, although one or two more recently published studies in bacteria have also been included. The field is moving so rapidly, however, that it becomes necessary to call a stop somewhere and start writing. Therefore, it's best to treat the patterns presented here as preliminary and then check their robustness against future work.

For the quantitative analyses, this chapter focuses on those studies sequencing entire genomes either directly, as in the phage work, or using next-generation high-throughput techniques, as is becoming increasingly common in bacteria and fungi. Other techniques for mutation discovery exist, but these are less comprehensive in their genome coverage. The methods, as well as their pros and cons, are discussed briefly in **Box 5.1.**

PATTERNS OF SUBSTITUTION DURING EXPERIMENTAL EVOLUTION

Chapter 2 presented two models of adaptation, one based on phenotypes and the other on DNA sequences. Both models made the prediction that most of the fitness gains during an adaptive walk in a constant environment are attributable to just a few beneficial mutations. The available evidence, though limited, is consistent with this prediction, but that could be because the non-sequencing methods for inferring beneficial substitutions miss genes of small effect and so underestimate true walk length. Genome-sequencing methods, on the other hand, might end up overestimating walk length either because of sequencing errors or because nonadaptive mutations get carried to fixation alongside beneficial ones (a process called hitchhiking).

How Many Mutations Are Typically Substituted during Adaptation?

It turns out that if we consider separately those studies that sequenced genomes (see Table 2.3) and compare them to nonsequencing studies, we find that adaptive walks are modestly longer in sequencing studies, but only by about one mutation. Recall that the studies in Table 2.3 were chosen because they showed evidence that the selection lines had reached a fitness plateau during the experiment and so we can be reasonably confident that the majority of adaptation has been accomplished. The median walk length for sequencing studies is 3, whereas that for the nonsequencing studies is 2.05. For the most part, the mutations uncovered in sequencing studies have the hallmarks of being adaptive, which we'll examine more closely later in this chapter. The important point is that adaptive walks, while longer than what is usually inferred based on nonsequencing methods, are not *that* much longer; they are still, by most accounts, fairly short. On a more technical point, it's crucial that future work not rely on nonsequencing methods to make inferences about adaptive walks unless their accuracy can be independently verified. Whole genome sequencing is now the gold standard for interpreting the genetics of adaptation in experimental evolution.

Table 5.1 Summary of the genetic changes identified through whole-genome sequencing of microbial selection experiments. SV stands for structural variant, a catch-all term for mutations above the level of single nucleotides such as insertions, deletions, duplications, and rearrangements.

Species	Genome size	Coding bases	GC bases	Number of genes	Citation	Selection environment	Lineage name
DNA VIRUSES							
phiX174	5386	5119	2411	11	Bull et al. 1997	High temperature, E. coli	C1
							C2
							C1+
						High temperature, Salmonella	S1
							S2
							S2
							S1+
phiX174	5386	5119	2411	11	Wichman et al. 1999	High temperature, Salmonella	Tx
							Id
phiX174	5386	5119	2411	11	Wichman et al. 2000	High temperature, E. coli	XC5
S13	5386	5119	2411	11			13C1
							13C2
							13C3
						High temperature, Salmonella	13S1
							13S2
							13S3
phiX174	5386	5119	2411	11	Holder and Bull 2001	High temperature, E. coli	n/a
G4	5577	5295	2548	11			G4-41.5
							G4-44
G4	5577	5295	2548	11	Rokyta, Abdo, and Wichman 2009	E. coli	ID12a
							ID12b
							ID8a
							ID8b
							NC6a
							NC6b
ID2	5486	5277	2510	11			ID2a
WA13	6068	5060	2720	11			NC28a
							WA13a
phiX174	5386	5119	2411	11	Wichman, Millstein, and Bull 2005	E. coli	n/a
MS2	3569	3246	1860	4	Betancourt 2009	Cold temperature, E. coli	M1
							M2
							M3

Generations	Large SV (>10 bp)	Small SV (2–10 bp)	Single nucleo-tide SV	Total SNPs	Non-genic SNPs	Genic SNPs	Synonymous SNPs	Proportion non-synonymous SNPs
1000	0	0	0	12	0	12	2	0.83
1000	0	0	0	12	0	12	1	0.92
2000	0	0	0	8	0	8	1	0.88
1000	0	0	0	15	1	14	0	1.00
1000	0	0	0	13	0	13	1	0.92
1000	0	0	0	13	0	13	2	0.85
2000	0	0	0	6	0	6	2	0.67
1000	1	0	0	14	0	14	0	1.00
1000	1	0	0	13	0	13	2	0.85
1000	0	0	0	7	0	7	1	0.86
1000	0	0	0	6	0	6	0	1.00
1000	0	0	0	8	0	8	1	0.88
1000	0	0	0	10	0	10	1	0.90
1000	0	0	0	13	0	13	0	1.00
1000	0	0	0	13	0	13	2	0.85
1000	0	0	0	8	0	8	0	1.00
498	0	0	0	2	0	2	0	1.00
664	0	0	0	3	0	3	1	0.67
1050	0	0	0	9	1	8	1	0.89
180	0	0	0	2	0	2	0	1.00
210	0	0	0	2	0	2	1	0.50
240	0	0	0	5	0	5	0	1.00
270	0	0	0	4	0	4	0	1.00
210	0	0	0	4	0	4	0	1.00
210	0	0	0	3	0	3	0	1.00
600	0	0	0	8	0	8	1	0.88
240	0	0	0	3	0	3	1	0.67
120	0	0	0	2	0	2	0	1.00
13000	0	1	1	101	9	46	46	0.00
50	0	0	0	4	0	4	0	1.00
60	0	0	0	3	0	3	0	1.00
70	0	0	0	4	0	4	0	1.00

(*continued*)

Table 5.1 (continued)

Species	Genome size	Coding bases	GC bases	Number of genes	Citation	Selection environment	Lineage name
RNA VIRUSES							
Vesicular stomatitis virus (VSV)	11161	10623	4658	5	Novella et al. 2004	Baby hamster kidney (BHK-21) cells	KA
							KB
							KC
							KD
						Sand fly (LL-5) cells	LB
							LD
						Alternating BHK-21/LL-5	LKA
							LKB
							LKC
							LKD
West Nile virus (WNV)	10962	10293	5597	10	Ciota et al. 2007	Mosquito (C6/36) cells	3356.1.1.1
St Louis encephalitis virus (SLEV)	10293	10293	5070	10			217.3.1.1
VSV	11161	10623	4658	5	Remold, Rambaut, and Turner 2008	Human cervical epithelial (HeLa) cells	H1
							H2
							H3
							H4
						Canine kidney epithelial (MDCK) cells	M1
							M2
							M3
							M4
						Alternating HeLa/MDCK	A1
							A2
							A3
							A4
Dengue virus (DENV)	10703	10167	4933	14	Vasilakis et al. 2009	Human hepatocyte-derived carcinoma cell (HuH-7)	P8-1407a
							P8-1407b
							IQT-1950a
							IQT-1950b
						Mosquito (C6/36) cells	P8-1407a
							P8-1407b
							IQT-1950a
							IQT-1950b
						Alternating HuH-7/C6/36	P8-1407a
							P8-1407b
							IQT-1950a
							IQT-1950b

Generations	Large SV (>10 bp)	Small SV (2–10 bp)	Single nucleo-tide SV	Total SNPs	Non-genic SNPs	Genic SNPs	Synonymous SNPs	Proportion non-synonymous SNPs
532	0	0	0	9	1	8	4	0.50
532	0	0	0	5	1	4	1	0.75
532	0	0	0	2	0	2	0	1.00
532	0	0	0	8	1	7	2	0.71
532	0	0	0	12	0	12	2	0.83
532	0	0	0	9	0	9	1	0.89
532	0	1	0	17	2	15	5	0.67
532	0	1	0	20	3	17	7	0.59
532	0	0	0	18	0	18	6	0.67
532	0	1	0	17	3	14	5	0.64
266	0	0	0	3	0	3	1	0.67
266	0	0	0	8	0	8	2	0.75
95	0	0	0	5	n/a	n/a	n/a	n/a
95	0	0	0	5	n/a	n/a	n/a	n/a
95	0	0	0	5	n/a	n/a	n/a	n/a
95	0	0	0	4	n/a	n/a	n/a	n/a
95	0	0	0	2	n/a	n/a	n/a	n/a
95	0	0	0	6	n/a	n/a	n/a	n/a
95	0	0	0	6	n/a	n/a	n/a	n/a
95	0	0	0	7	n/a	n/a	n/a	n/a
95	0	0	0	5	n/a	n/a	n/a	n/a
95	0	0	0	8	n/a	n/a	n/a	n/a
95	0	0	1	9	n/a	n/a	n/a	n/a
95	0	0	0	5	n/a	n/a	n/a	n/a
66	0	0	0	4	0	4	3	0.25
66	0	0	0	2	0	2	0	1.00
66	0	0	0	3	0	3	0	1.00
66	0	0	0	2	0	2	0	1.00
66	0	0	0	2	1	1	1	0.00
66	0	0	0	3	1	2	1	0.50
66	0	0	0	1	0	1	0	1.00
66	0	0	0	1	0	1	0	1.00
133	0	0	0	3	1	2	1	0.50
133	0	0	0	2	1	1	1	0.00
133	0	0	0	1	0	1	1	0.00
133	0	0	0	2	0	2	1	0.50

(*continued*)

Table 5.1 (continued)

Species	Genome size	Coding bases	GC bases	Number of genes	Citation	Selection environment	Lineage name
BACTERIA							
Escherichia coli	4639675	3992744	2356477	4497	Shendure et al. 2005	syntrophic symbiosis	n/a
	4629812	4040395	2350528	4312	Barrick et al. 2009	glucose limitation	Ara-1 40K
	4639675	3992744	2356477	4497	Conrad et al. 2009	lactate	LactA
							LactB
							LactC
							LactD
							LactE
							LactF
							LactG
							LactH
							LactI
							LactJ
							LactK
					Atsumi et al. 2010	isobutanol	SA481
					Lee and Palsson 2010	propanediol	eBOP12
	4578159	4194000	2325705	4255	Maharjan et al. 2010	glucose limitation	BW4001
					Wang et al. 2010	phosphate limitation	4218
							4223
							4227
							4236
							4239
	4639675	3992744	2356477	4497	Minty et al. 2011	isobutanol	G3.266.7
							X3.5
					Charusanti et al. 2010	glucose	pgi_gluc1
							pgi_gluc7
							pgi_gluc10
Myxococcus xanthus	9139763	8298200	6296496	7454	Velicer et al. 2006	social cheating	PX
Pseudomonas aeruginosa	6537648	5864198	4334461	5977	Wong, Rodrigue, and Kassen 2012*	synthetic cystic fibrosis medium (SCFM)	Pa14
						SCFM + ciprofloxacin	
						SCFM+mucin	
						SCFM+ciprofloxacin+mucin	
Pseudomonas fluorescens	7147633	6316305	4293390	6492	Beaumont et al. 2009	Artificial, alternating morphotype selection	1B4

Generations	Large SV (>10 bp)	Small SV (2–10 bp)	Single nucleo-tide SV	Total SNPs	Non-genic SNPs	Genic SNPs	Synonymous SNPs	Proportion non-synonymous SNPs
200	0	1	0	2	1	1	0	1.00
20000	11	0	5	29	7	22	0	1.00
1100	1	0	0	3	0	3	1	0.67
1100	1	0	1	0	0	0	0	n/a
1100	1	0	0	2	0	2	0	1.00
1100	1	0	0	6	0	6	1	0.83
1100	2	0	0	6	0	6	0	1.00
750	2	0	0	3	0	3	0	1.00
750	1	0	2	5	1	4	2	0.50
750	1	1	2	2	0	2	0	1.00
750	1	0	1	1	0	1	0	1.00
750	1	0	1	4	1	3	1	0.67
750	1	1	0	5	1	4	0	1.00
299	26	0	0	1	0	1	n/a	n/a
700	1	0	2	3	1	2	1	0.50
89	0	0	0	1	0	1	0	1.00
127	0	0	0	3	2	1	0	1.00
127	1	0	0	1	0	1	0	1.00
127	0	0	1	2	1	1	0	1.00
127	0	0	0	3	1	2	0	1.00
127	1	0	0	5	1	4	0	1.00
266	4	2	1	1	0	1	0	1.00
500	2	2	3	4	0	4	1	0.75
800	0	0	0	5	0	5	0	1.00
800	1	0	0	3	1	2	0	1.00
800	0	1	0	2	0	2	0	1.00
1060	0	0	1	14	3	11	1	0.91
47.5	0	0	0	0.17	0	0.17	0.08	0.53
47.5	0.08	0.08	0.5	2.08	0	1.42	0	1.00
47.5	0	0	0	0.83	0.08	0.75	0	1.00
47.5	0.3	1	0.2	2.7	0	1.2	0	1.00
n/a	1	0	3	5	0	5	0	1.00

(*continued*)

Table 5.1 (continued)

Species	Genome size	Coding bases	GC bases	Number of genes	Citation	Selection environment	Lineage name
FUNGI							
Saccharomyces cerevisiae	12156600	8823707	4637520	6273	Anderson et al. 2010	High salt	S2
							S6
						glucose limitation	M8
					Araya et al. 2010	sulfate limitation	DBY11331
					Kvitek and Sherlock 2011	glucose limitation	M5Yellow
							M4Red
							M1Green
Pichia stipitis	1540000	n/a	n/a	5841	Smith et al. 2008	xylose fermentation	Shi21

* Average SNPs across 12 independently evolved genotypes per treatment

Rates of Substitution

Because adaptive walks involve just a few substitutions, the rates of substitution likely can be quite fast. We can readily estimate substitution rates from the data provided in Table 5.1. For easier comparison across studies, in this table the single nucleotide polymorphism (SNP) substitution rate is calculated every 100 generations per 1000 base pairs (bp). The choice of scaling here is effectively arbitrary, although it's somewhat justified because 100-generation intervals are often sampled in many selection experiments, and mutational landscape models often use 1000 bp as the length of a "typical" gene (Orr 2002). The focus is on SNPs because they are most effectively detected by high-throughput NGS.

Figure 5.2a shows the results. Two things immediately pop out. The first is that compared to bacteria and fungi, viruses have orders of magnitude higher rates of substitution. This difference no doubt reflects the higher mutation rate of phages compared to bacteria and fungi. The second is that, for all three groups, substitution rates decline as the duration of the experiment increases. This finding probably reflects the decreased supply and/or effect size of beneficial mutations as populations approach adaptive peaks.

When a population sits on or very near an adaptive peak, all new mutations are likely to be either neutral or deleterious. Recall that when mutations are neutral, the rate of substitution *is* the mutation rate. We might therefore expect substitution rates over the very long term to approach the underlying mutation rate. This is, in fact, what we see in **Figure 5.2b** for experiments

Generations	Large SV (>10 bp)	Small SV (2–10 bp)	Single nucleo-tide SV	Total SNPs	Non-genic SNPs	Genic SNPs	Synonymous SNPs	Proportion non-synonymous SNPs
500	1	0	0	5	1	4	0	1.00
500	1	0	0	4	2	2	0	1.00
500	0	0	0	6	0	6	1	0.83
188	1	0	0	4	1	3	0	1.00
385	2	0	0	5	0	5	1	0.80
266	1	0	1	3	1	2	0	1.00
56	0	0	0	1	0	1	0	1.00
n/a	0	0	0	13	4	9	0	1.00

with *E. coli,* where substitution rates (gray dots) are initially high and then approach that of published mutation rates for genes not under selection as time goes on. For the phage ΦX174, on the other hand, substitution and mutation rates appear to be roughly the same no matter how long the experiment continues (black dots). Mutation rate estimates are from Lynch (2010).

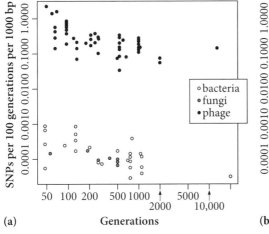

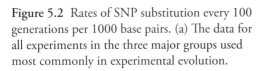

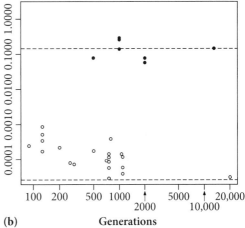

Figure 5.2 Rates of SNP substitution every 100 generations per 1000 base pairs. (a) The data for all experiments in the three major groups used most commonly in experimental evolution.

(b) The same data as in (a), but for the phage ΦX174 and the bacterium *E. coli* only. Dashed lines represent the estimated (scaled) mutation rate for each species provided by Lynch (2010).

Box 5.1

Methods for Mutation Discovery in Experimental Evolution

1. *Direct sequencing* Small genomes can often be sequenced directly using the Sanger chain-termination method. The basic idea is to sequence the genome by amplifying smaller regions of a few hundred base pairs using the polymerase chain reaction (PCR) and then assembling the different fragments into a single genome. Comparison against the ancestral genome allows mutation identification. The advantage of this technique is that the sequence data is of high quality, but the need to manually construct many sets of primers and process the DNA for sequencing greatly limits its use to very small genomes, typically those on the order of a few thousand base pairs. The method is therefore appropriate for some phage genomes but not much else.

2. *Candidate gene sequencing* Species with large genomes, which basically means anything above the size of a few thousand base pairs, cannot reasonably be sequenced in their entirety using direct sequencing methods. An alternative approach is candidate gene sequencing—a method that relies on prior knowledge, and a good amount of luck—to screen the potential genetic targets of selection for mutations. Identified targets, which are often genes associated with various physiological and metabolic processes, are then sequenced directly through conventional, primer-based PCR approaches. The method is not comprehensive, and the need to rely on prior knowledge restricts its use to all but the best-characterized model organisms such as *E. coli* and yeast. Nonetheless, this approach has been successful at uncovering mutations in at least some analyzed loci (Cooper, Rozen, and Lenski 2003; Crozat et al. 2010, 2005; Kinnersley, Holben, and Rosenzweig 2009; Notley-McRobb and Ferenci 1999a, 1999b; Ostrowski, Woods, and Lenski 2008; Treves, Manning, and Adams 1998; Woods et al. 2006). These mutations rarely account completely for the observed changes in fitness in the evolved lines.

3. *High-throughput sequencing technologies* It is now technically and financially feasible for individual laboratories to obtain the entire (or very close to the entire) genome sequence of multiple strains from large numbers of evolved

To understand these contrasting results, remember that if a population is near an adaptive peak, then by definition no beneficial mutations are available to it; the only mutations being fixed are neutral, or nearly so. This is not what happens in these experiments, however. Both the *E. coli* experiment and the phage experiment yield abundant evidence that the mutations being substituted are adaptive: they tend to be non-synonymous, occur in open reading populations using so-called next-generation sequencing (NGS) technologies (Brockhurst, Colegrave, and Rozen 2011; Dettman et al. 2012). The basic strategy is to detect mutations by searching for polymorphisms between ancestral and evolved genomes. Although different sequencing platforms use different underlying technologies, they all work in essentially the same way: generate an extremely large number of short sequence reads, which are pieced together using a previously sequenced reference genome as a guide. Processing the enormous amounts of raw sequence data generated from high-throughput sequencing is not trivial; it requires specialized bioinformatics tools to manage and sort through the data to identify genuine mutations from sequencing errors. Most labs have developed their own custom-built sequencing pipelines to process the raw data to identify and annotate potential mutations. A sample pipeline developed and built from freely available software is provided in Dettman et al. (2012). A word of caution: identifying mutations is still something of an art, because the parameters used to process the data and the strictness of the values used to detect polymorphism can affect the rates of false positives or negatives. It's therefore worth treating the list of mutations returned as hypotheses and, wherever possible, confirming that they are genuine mutations through direct sequencing using PCR-based approaches that have a much lower error rate. It is also good practice to confirm that these mutations are adaptive by measuring their fitness in competitive assays against the ancestor.

4. *Microarray-based mutation detection* This method detects mutations by using DNA–DNA hybridization patterns between a reference and an evolved genome. Since hybridization strength is quantitatively sensitive to the number of mismatches in the DNA, mutations can be detected as regions where hybridization is weaker compared to non-mutated DNA (Gresham et al. 2006, 2010). However, the precise position and type of mismatch, as well as the exclusion of false positives, must be confirmed through PCR-based sequencing (Gresham et al. 2008; Herring et al. 2006). A further drawback of microarray-based methods is that they have been shown to miss up to 30 percent of the mutations found through NGS (Araya et al. 2010; Gresham et al. 2008; Herring and Palsson 2007; Herring et al. 2006; Kao and Sherlock 2008; Kvitek and Sherlock 2011). Even so, a potential advantage of this technique is that it can identify large regions of structural variation such as insertions, duplications, or deletions that can be missed by high-throughput techniques due to their reliance on short sequence reads.

frames, and often occur in multiple, independently evolving populations—consistent with the idea that these mutations are under strong selection. So these populations are still adapting, although at rates that are hard to distinguish from background mutation rates.

These populations thus seem to be in a state of "perpetual" adaptation, either because the approach to a fitness optimum takes an extremely long time or because evolutionary changes in the composition of the population cause changes in the biotic environment that prevent a population from ever reaching a fitness peak. A number of microbial studies have noted situations where the biotic environment can change in ways that seem to lead to such perpetual adaptation. The changes occur both in the absence of antagonistic species interactions (Friesen et al. 2004; Meyer et al. 2011; Paquin and Adams 1983; Rainey and Rainey 2003; Spencer et al. 2008) and in their presence (Buckling and Rainey 2002a; Paterson et al. 2010). If this is a general effect, then it's possible that adaptive walks, once started, may never actually finish.

Temporal Patterns in Genomic Evolution

Rates of fitness increase tend to be faster early in a selection experiment and slower later on, once the population has adapted (Chapter 2). Rates of substitution show a similar pattern. Ultimately, this pattern occurs because the selection coefficients associated with the first substitutions confer larger fitness increases than later ones. It's important here to distinguish between a large-effect substitution and a large-effect mutation, because the two are not always equivalent. A large-effect substitution may be caused by a single beneficial mutation with a large fitness effect (Chou et al. 2011; Khan et al. 2011), or it may be caused by multiple beneficial mutations that have arisen in the same genome and are having a large combined effect (Conrad et al. 2009; Lee and Palsson 2010). This latter phenomenon, also known as the leapfrog effect (Desai and Fisher 2007), can occur only in populations that are large enough for independently arising beneficial mutations to compete for fixation.

The Genomic Geography of Mutation

How are mutations distributed in an evolved genome? In non-recombining populations, the answer depends on the relative strength of selection and drift. Drift will cause mutations to be substituted randomly across the genome, while selection will cause mutations to be substituted in regions that are important for fitness. One expects selection to be strong in most microbial experiments, since population sizes are large and the founding genotypes are often, by design, maladapted. Thus we expect to see SNPs overrepresented in regions of the genome that code for a gene product (genic region) and under-represented in regions that are noncoding (intergenic). We should also see that most of the substituted SNPs are non-synonymous—they change the amino acid in a protein—because they are more often under selection than synonymous SNPs that do not result in an amino acid replacement.

Figure 5.3 gives us a sense of where SNPs tend to be found in microbial selection experiments in viruses, bacteria, and single-celled fungi. The left-

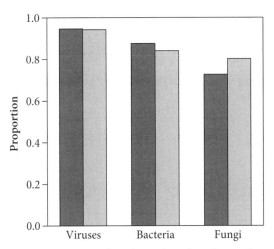

Figure 5.3 The genomic geography of single nucleotide polymorphisms (SNPs). Bars show the fraction of the genome that is open reading frame (dark gray) and the SNPs that occur in open reading frames (light gray).

hand bars represent the average fraction of the genome that is coding. Most of a microbial genome is coding, so this fraction is very high, even in single-celled eukaryotes like yeast. The right-hand bars represent the proportion of the genome that contains SNPs in open reading frames. Again, this fraction is high for all groups. It's difficult to detect any real signal in the viral or bacterial data, but that may be thanks to the density of open reading frames in these genomes: viruses are on average 95 percent coding, and bacteria are 88 percent. Fungi tend to have SNPs overrepresented in coding regions, relative to their frequency in the genome. In fungi, there are more noncoding DNA and a lower gene density (73 percent), but the fraction of SNPs occurring in coding genes in fungi is high (80 percent). The fungal data is thus consistent with selection operating on functional coding DNA during adaptation.

We can also calculate the fraction of SNPs in coding regions that are non-synonymous and so are likely to be under selection. Most SNPs are non-synonymous, and Table 5.1 shows that many experiments find exclusively non-synonymous mutations. The mean fraction of SNPs that are non-synonymous for viruses is 77 percent, for bacteria 91 percent, and for fungi 95 percent. The exact proportion of non-synonymous sites in a genome varies from species to species due to factors like codon bias (the tendency for certain codons to be overrepresented in a genome compared to others), but it is often between 75 percent and 80 percent, which means there is good evidence from bacteria and fungi that most SNPs are adaptive. The lower fraction of non-synonymous substitutions in viruses likely reflects their overall higher rates of mutation. A synonymous mutation may end up hitchhiking to high frequency just because it happens to arise in the background of a beneficial mutation.

There is an alternative explanation, however—that synonymous SNPs may not always be neutral. Viruses, for example, contain many overlapping reading frames, where the same DNA sequence codes for more than one gene, so a mutation can be synonymous for one gene but not for the other. Synonymous mutations may also affect transcription and translation in a variety of ways that can in principle lead to their having significant fitness effects (Plotkin and Kudla 2011). In fact, Bailey, Hinz and Kassen (2014) recently have found two synonymous mutations in the same gene that increase fitness in a population of *Pseudomonas fluorescens* that had evolved for 1000 generations in a minimal glucose medium. The mutations occur in the permease gene of an ABC transporter that is necessary for bringing glucose across the inner membrane of the cell. These mutations have selection coefficients that are comparable to any of the non-synonymous mutations that arose in the same experiment. The mechanism by which they confer their fitness advantage seems to involve increasing rates of gene expression of the operon that the gene is a part of, although the details of how this occurs remain elusive. It is not yet known how common this phenomenon is in natural systems, but if it is fairly frequent, then it will be necessary to rethink the basic assumption that synonymous mutations are always neutral.

Adaptive versus Nonadaptive Substitutions

The data reviewed so far—higher rates of substitution compared to the neutral expectation, the predominance of non-synonymous substitutions, and the tendency for observed SNPs to occur in open reading frames—suggests strongly that the mutations being detected through whole genome sequencing are contributing to adaptation. Yet it is not uncommon to see mutations that have no noticeable beneficial effects (Anderson et al. 2010; Barrick et al. 2009; Conrad et al. 2009; Lee and Palsson 2010; Minty et al. 2011; Wang et al. 2010), and sometimes even detrimental effects (Conrad et al. 2009; Crozat et al. 2010; Kvitek and Sherlock 2011), when tested in the ancestral genetic background. Why?

The simplest explanation is that these mutations are fixing because they hitchhike alongside beneficial mutations that are themselves under selection. Because the experiments are all done under conditions where reproduction is completely asexual, hitchhiking is extremely likely. An alternate possibility is that fitness was not measured appropriately. The best measure of fitness comes from direct, head-to-head competition experiments, but it's not uncommon to see only a single component of fitness like growth rate or stationary phase density reported in the literature. It's also important to have the right competitor strain to test fitness against. In most experiments, fitness is estimated relative to the ancestor that was used to found the selection lines. But if the genetic environment in which selection is acting changes over the course of the experiment, then this approach can give misleading results. One example of this phenomenon has been observed in phage, where a changing genetic environment apparently generated continued selection (Wichman, Millstein, and Bull 2005). Other examples include the emergence of frequency-dependent fitness relationships (Friesen et al. 2004; Lang, Botstein, and Desai 2011; Rainey and Travisano 1998) and where sequential substitutions have non-transitive fitness effects relative to the ancestor (Paquin and Adams 1983). The more relevant fitness measure in these cases is when the genotype of interest is competed against the immediate genetic background that it evolved from. Aside from the one phage study by Wichman, Millstein, and Bull (2005), no systematic efforts have been made to tie whole genome sequencing data to fitness in these more complicated situations.

A third possibility is epistasis, where the fitness of a mutation depends on other mutations present in the genome. Epistasis appears to be quite common in experimental evolution, but not all forms of epistasis lead to apparently nonadaptive mutations. To see this, first we need to be clear on the different kinds of epistasis that exist among beneficial mutations.

Negative epistasis is the condition when mutations have larger beneficial effects singly than they do in combination. It cannot, by definition, explain apparently nonadaptive mutations.

The opposite of negative epistasis is positive epistasis, which occurs when the fitness effect of mutations in combination is greater than would be expected from the sum of their independent effects. This kind of epistasis can generate apparently nonadaptive mutations because a mutation on its own may be neutral, but in combination with a second mutation may be beneficial. Positive epistasis has been observed in a number of experiments (Anderson et al. 2010; Applebee, Herrgård, and Palsson 2008; Conrad et al. 2009; Herring et al. 2006; Lee and Palsson 2010; Minty et al. 2011). Lee and Palsson (2010) provide a particularly striking example. They showed how two mutations, *IS5* and *fucO*, from an evolved line of *E. coli* adapted to propanediol (on which the ancestor could barely grow) were neutral in the ancestral genome when tested singly but accounted for over half of the adaptive fitness increases when tested in combination.

Finally there is sign epistasis, which occurs when the sign of the fitness value of a mutation changes from negative to positive (or vice versa) depending on the presence of other mutations. This form of epistasis can also lead to apparently nonadaptive mutations, because two mutations that on their own decrease fitness will in combination increase it. An example of this form comes from lactate-adapted populations of *E. coli* (Conrad et al. 2009). Four independently evolved lines all had a mutation in the cell envelope gene *kdtA* that, when tested in the ancestral genetic background, actually killed the cell. Cell death occurred because the mutation made the cell auxotrophic for some key amino acids, meaning that the cell could not synthesize the amino acids on its own. When the *kdtA* mutation was tested together with co-acquired mutations, the auxotrophy requirement was relieved and the combined mutations were significantly beneficial.

The Importance of Epistasis for Genomic Trajectories

The foregoing discussion underscores the prevalence of epistasis among beneficial mutations. Epistasis, and in particular sign epistasis, has an important impact on evolving populations because it can profoundly affect the trajectory of evolution. A classic example comes from the work of Weinreich et al. (2006), who showed that among 120 possible genetic routes by which *E. coli* can become resistant to the antibiotic cefotaxime (representing all possible pathways among five distinct mutations), only 18 represent viable pathways in the sense that each step along the pathway involved an increase in the level of resistance. The other 102 pathways, by contrast, all involved decreases in resistance at some step along the trajectory and so would be evolutionarily inaccessible due to sign epistasis.

Kvitek and Sherlock (2011) provide a second, even more striking, example of how sign epistasis can lead to different evolutionary outcomes. They sequenced genomes from subpopulations of a single glucose-limited yeast population that had evolved for just under 500 generations and found two mutations (*MTH1* and *HXT6/7*) that each, on their own, increased fitness by increasing expression of a glucose transporter. However, the two mutations were never found in the same genome. And, when they were tested together in the ancestral background, the double mutant was actually less fit than the ancestor. The mechanistic cause of reciprocal sign epistasis is not clear, but it may be the result of an overabundance of hexose transporter proteins that could lead to an inefficient allocation of resources within the cell. Regardless, the strong reciprocal sign epistasis between these two mutations caused adaptive evolution to take markedly different genomic pathways.

PARALLEL AND CONVERGENT EVOLUTION

Do the same genomic changes happen repeatedly when evolution is allowed to run its course again? Stephen Jay Gould famously called this process "rewinding the tape of life" and, for him, it was a thought experiment. With experimental evolution, researchers now can actually do the experiments Gould thought impossible. We inoculate multiple independent populations of the same starting genotype and monitor how evolution unfolds in each one. In this way, the tape of life can be replayed as many times as space in the laboratory permits and the experimenter's patience allows.

Definitions

The concepts of parallel and convergent evolution have been around for some time, but only recently has it become possible to study them at the genomic level. The basic idea is that finding the same mutations fixed in independently evolved populations is strong evidence that the mutations are adaptive, because it's highly unlikely that exactly the same mutations would be substituted by chance alone. The repeated fixation of the same mutations is called parallel evolution when the founding genotype is the same for all populations; it is called convergent evolution if the founding genotypes are different. It's important to remember that, as tempting as it is to attribute such repeated fixations to selection, it is always possible that fixation of the same genetic changes occurs by chance alone—especially in very small genomes such as phage, where the number of potential variable sites can be quite limited. It is thus always necessary to distinguish repeatability due to chance from that due to selection in experimental design and analysis. One way of doing this is to examine the repeatable changes themselves for signatures of adaptive substitution, such as how rapidly they increase or whether they are non-synonymous. An alternative approach is to develop more quantitative criteria based on the number of variable sites across the genome, but to the best of my knowledge this has not been done. In the interim being conservative in assessing repeatability seems appropriate. In phage experiments, for example, it is common to assign a substitution as repeatable only if it is observed in three or more independently evolving lines.

The Hierarchy of Repeatability

The frequency of parallel evolution should depend on the level of biological organization being examined. When examining fitness only, adaptive evolution is usually highly repeatable, provided there is appropriate genetic variation for fitness and selection is strong enough to overcome drift and migration. But things get more complicated when we examine lower levels of biological organization. In most organisms and in most environments, there are probably many different combinations of phenotypes, and even more combinations of genotypes, that can generate high fitness. The degree to which adaptive evolution is repeatable should therefore decrease as we go lower down the biological hierarchy: repeatability will be highest at the phenotypic level, lower at the level of genes themselves, and lowest at the level of individual nucleotides.

Gresham et al. (2008) have explored the degree of parallelism at different levels of biological organization. They allowed yeast lines to adapt to one of three kinds of nutrient limitation—glucose, phosphate, or sulfate—for approximately 200 generations and then examined the patterns of response in cell physiology, gene expression, and nucleotide changes. Replicate lines within an environment responded in parallel to selection. Cell volume, for example, increased in seven of eight replicate cultures evolved in phosphate, and residual glucose concentrations in the media were vastly decreased in all glucose-evolved populations. At the level of gene expression, responses were somewhat more variable and depended on the selection environment. In sulfur, all replicate lines showed a similar gene expression phenotype; but in glucose and phosphate, there was substantially more variation in gene expression—each environment included up to three distinct gene expression profiles. At the nucleotide level things were even more variable. No loci or nucleotides were shared among replicates in glucose or phosphate, and only one locus was shared in replicate sulfate lines, although with different nucleotide changes.

A striking feature of this study is that the degree of parallelism seems to depend on the environment. Lineages evolving under sulfate limitation displayed more parallelism than those evolving under phosphate or glucose limitation. This is not an unusual result. In work with the soil microbe *Pseudomonas fluorescens*, for example, investigators have seen parallel and convergent evolution toward a common metabolic profile following selection in minimal media with glucose (Melnyk and Kassen 2011). When glucose is replaced with xylose, however, at least two markedly different metabolic profiles evolved, irrespective of founding genotype. Similar results have been observed in *E. coli* where populations evolved in parallel in the same environment show quite different gene expression profiles (Fong, Joyce, and Palsson 2005; Maharjan et al. 2006). It's tempting to interpret these results to mean that the underlying adaptive landscape—how genotype or phenotype maps to fitness—is more rugged in some environments than others, presumably because of sign epistasis between alternative combinations of alleles. However, such an interpretation awaits direct experimental confirmation.

A Theory of Repeatability

Quantitative statements about the probability of parallel evolution at the locus and nucleotide levels can be made using the models of adaptation introduced in Chapter 2. Among the more compelling predictions to emerge is that the probability of parallel evolution should increase as a population approaches a fitness optimum, simply because the number of beneficial mutations available to selection goes down (Chevin, Martin, and Lenormand 2010; Orr 2005). In fact, a startlingly simple prediction emerges from the mutational landscape model: the probability of parallel evolution under natural selection is approximately $2/(n+1)$, where n is the number of beneficial mutations available to a given sequence (Orr 2005).

Other predictions can be made, but it is less clear how robust they are. Chevin, Martin, and Lenormand (2010) showed that parallel evolution becomes more likely when fewer independent traits are affected by a mutation, as can happen if there are strong genetic covariances among traits, thus reducing the effective number of possible routes that can be taken. Presumably, clonal competition increases the probability of parallel evolution even more by constraining the mutations that fix to be those with the largest selective advantages, but few attempts have been made to quantify this effect. The ruggedness of the underlying adaptive landscape might also affect parallel evolution. The probability of parallel evolution might be lower in a rugged than in a smooth landscape, because multiple adaptive peaks are presumably available to a genotype in the former but not in the latter.

Data on Repeatability

Strong quantitative tests of these predictions are not usually possible, because doing so requires knowing quantities like n, the degree of pleiotropy among mutations, or the ruggedness of the adaptive landscape. These values are hard, if not impossible, to quantify. Still, there is some data on the relative frequency of locus- and nucleotide-level parallel evolution. Woods et al. (2006) contrasted the extent of parallel evolution at four candidate loci that contained mutations following 20,000 generations of selection with that from 36 randomly selected loci. Across 12 independently evolving populations, parallel evolution was much more common in the four candidate loci than in the 36 randomly selected loci, and two genes (*nadR, pykF*) had mutations in all 11 other replicate lines. Interestingly, though, parallel evolution at the nucleotide level was very rare. Of two nucleotide substitutions found in common among independent lines, one (in *pykF*) was found in three populations and the other (in *nadR*) was found in two.

These results are not extraordinary, at least for bacteria and yeast. A single locus (*glpK*) was mutated in five glycerol-adapted lines of *E. coli* from one study (Herring et al. 2006) and, in yeast, amplifications of glucose transporter genes are common in glucose-limited environments (Brown, Todd, and Rosenzweig 1998; Dunham et al. 2002; Gresham et al. 2008; Kvitek and Sherlock 2011). Adaptation to the antifungal agent nystatin in yeast was highly repeatable at the level of the gene targeted: 33 out of 35 lines contained mutations in just two genes, *ERG6* and *ERG3,* both of which are involved in ergosterol synthesis (Gerstein, Lo, and Otto 2012). But it seems a stretch to say that parallel evolution is common. At the nucleotide level, parallel evolution is all but absent. Among 11 replicate lines in one study, for example, 72 percent of target loci were unique to a single line (Conrad et al. 2009). Tenaillon et al. (2012) saw comparable results when *E. coli* is allowed to adapt for

2000 generations to high temperatures. In the yeast study by Gerstein and coworkers, 7 in 19 lineages with mutations in *ERG6* were unique (~37 percent); in *ERG3,* the ratio was 11 in 14 (79 percent). Microarray-based studies give a similar picture (Gresham et al. 2008; Herring et al. 2006). Thus, even if parallel evolution at the gene level is sometimes common, it is not typically underlain by comparable degrees of repeatability at the nucleotide level, at least in bacteria and yeast.

In phages, things are quite different. In reviewing the many experiments that have used the microvirid phage ΦX174, Wichman and Brown (2010) found that, for any pair of evolved lines, about 50 percent of substitutions at the nucleotide level would be common to both, even after accounting for the possibility that some substitutions in this small genome could be the same by chance. This extremely high level of parallel evolution is probably due to two unique features of phages. First, phages have higher mutation rates and their genomes are smaller compared to yeast and bacteria, making it easier for phages to sample the entire range of mutations available. Second, phage genomes are much more compact and have genes that often share open reading frames. This situation may impose additional constraints on which mutations are likely to be beneficial, as it requires that a mutation that is beneficial for one gene does not severely compromise function in the other gene that shares the same site. Presumably the number of possible beneficial mutations goes up as genome size increases, which should lead to lower rates of parallel evolution. It would be interesting to contrast the rate of parallel evolution in a range of phage genomes that vary in size and gene number: T-odd and T-even phages have genomes that are two orders of magnitude larger than the microvirid phages, for example, and so might be expected to exhibit less parallel evolution than, say, ΦX174 (Bull et al. 2004).

THE GENOMIC ARCHITECTURE
OF ADAPTATION

So far, we have been considering the broad patterns of genomic changes that occur during adaptive evolution. Let's turn now to the genes themselves and ask whether particular kinds of genes, or genetic architectures, are more likely to be involved in adaptation than others.

At first, answering this question seems like an incredibly daunting task—especially for evolutionary biologists, who have come to see genes as abstract, generic units of inheritance. If there is one thing we have learned from the last 30 to 40 years of molecular biology, it is that the structure of genes and genomes is outstandingly complicated, involving many layers of regulation and interaction that we still don't fully understand. Yet the development of a predictive theory of evolution needs to go beyond simple statements like "Fitness will increase in a novel environment." We need to be able to say more about what the likely targets of selection are going to be. Let's review the various attempts that have been made to develop predictions about the targets of selection in evolving populations and discuss the most compelling—and persistent—hypotheses put forward to explain which genetic targets are the most likely to be involved in adaptive evolution.

Candidate Gene Approaches

The candidate gene approach relies on prior knowledge of the physiological and metabolic pathways involved in dealing with a particular stressor or agent of selection to make educated guesses about the genes likely to be involved in adaptation. In cases involving extremely strong selection, like the evolution of antibiotic or pesticide resistance, the biochemical targets of the drugs being used are often known in advance, making this approach quite successful. For example, quinolone antibiotics were first introduced in the 1960s to treat urinary tract infections and were explicitly designed to interfere with the molecular machinery involved in winding and unwinding DNA during replication. It's no surprise, then, that the most common, although by no means exclusive, mutations conferring resistance to quinolones are often in the very targets they were meant to inhibit—DNA gyrases and topoisomerases (Wong and Kassen 2011).

When more genetic targets are involved and selection is weaker, the candidate gene approach has been less successful. Mutations are often found in some loci, but they rarely account completely for the fitness changes observed during experimental evolution (Cooper, Rozen, and Lenski 2003; Crozat et al. 2010, 2005; Kinnersley, Holben, and Rosenzweig 2009; Notley-McRobb and Ferenci 1999a, 1999b; Ostrowski, Woods, and Lenski 2008; Treves, Manning, and Adams 1998; Woods et al. 2006). Knowledge of the genetic components of pathways is thus giving us only part of the story; when we rely solely on prior knowledge, we miss some of the mutations contributing to fitness.

In fact, whole genome sequencing studies do uncover mutations that have adaptive value but whose mechanistic function, at least in the context of the phenotype being examined, is not understood. The recovery of social cooperation in the bacterium *Myxococcus xanthus,* for example, was caused by a single point mutation in the promoter of a GNAT family acetyltransferase gene whose role in physiology and development had not been characterized (Velicer et al. 2006). Another example is the evolution of rapid colony morph switching following intense artificial selection for switching ability in *Pseudomonas fluorescens.* This phenotype arose due to a nucleotide change in the large subunit of carbamoyl-phosphate synthetase (Beaumont et al. 2009). How and why these mutations work to produce their respective phenotypes is completely unknown.

Cis-*Regulatory Changes*

Developmental biologists have argued that the bulk of evolutionarily important phenotypic variation (the variation that distinguishes major body plans in vertebrates) in plants and animals may be attributable to changes in gene regulation rather than to the gene products themselves (Stern and Orgogozo 2009). This idea derives primarily from the study of interspecies differences in morphology and has not been without controversy (Hoekstra and Coyne 2007). Nevertheless, it has been hugely influential and deserves closer examination.

To understand the argument, it's first necessary to have a primer on gene expression. Gene expression involves two components: the gene product itself—the protein or RNA coded for by the sequence of base pairs that gets transcribed—and the regulatory region of DNA, where transcription factors bind to control when and where the gene is expressed. Because transcription factor-binding regions are usually upstream of the genes that actually code for a protein, they are called *cis*-regulatory regions, and the protein-coding genes are often referred to as structural genes. Adaptation can proceed through mutations in either structural or regulatory sites.

Developmental biologists have argued that the regulatory mutations are the most important in adaptation. The reason for this claim is that regulatory mutations are expected to be less pleiotropic in their effects than are those in protein-coding regions. Less pleiotropic mutations are more likely to be beneficial, at least in Fisher's geometric model. It's harder, after all, to improve many linked traits simultaneously than it is to improve just one.

The crucial issues then are, first, whether regulatory mutations are more or less pleiotropic than structural mutations; and, second, whether such reduced pleiotropy does actually make them more likely to be beneficial. Advocates of the *cis*-regulatory model argue that a mutation affecting a coding region can have manifold results: it affects the protein being produced as well as all other proteins that the new protein interacts with. If that protein is itself a transcription factor, for example, the mutation will affect every gene it regulates. By contrast, a mutation in an upstream promoter region affects only the expression of that one gene. Thus mutations in structural genes, it is argued, are more likely to be highly pleiotropic—and so less likely to be beneficial—in their effects than mutations in regulatory binding sites.

In principle, this prediction could be tested simply by tallying the number of SNPs uncovered in experimental evolution in *cis*-regulatory and protein-coding DNA. In practice, this test is not usually possible because the many layers of regulation make it difficult to identify which regions are strictly regulatory and which are protein-coding. In addition, microbes contain precious little noncoding DNA, further complicating the distinction. Nevertheless, because the bulk of SNPs uncovered in experimental evolution are located in coding regions (as described earlier), we know that, in the strictest sense, the prediction that most genetic targets of adaptive evolution are in *cis*-regulatory sites is simply wrong for microbes.

Increased Gene Expression

Mutations that affect gene expression may be important for adaptive evolution, even if they are not specifically *cis*-regulatory. When faced with a novel environment or resource, most organisms have some crude ability at least to get by with the metabolic machinery already in place. One of the easiest ways of increasing fitness in a novel environment may therefore be to make more of what you already have, even if what you have isn't the best it could be for supporting growth. Increases in the amount of gene product can thus allow an organism to persist and grow in a novel environment. Most of the early genetic changes involved in adaptation may therefore involve increased gene expression levels via mutations in regulatory genes. Once the population is able to persist, natural selection can modify existing structural genes to create new enzymes. This is the exaptation-amplification-diversification (EAD) model for the evolution of novel gene functions (Bergthorsson, Andersson, and Roth 2007), and it leads to the prediction that a regular sequence of events should be involved in adaptation: regulatory mutations leading to increased gene expression come first, followed by mutations to structural genes later.

Testing the idea that regulatory mutations should be common is challenging when using whole genome sequencing data, because it's rarely possible to identify convincingly that a gene is regulatory or non-regulatory. Given the highly interdependent nature of cellular networks, even genes that appear to be non-regulatory can have regulatory functions. An opportunity exists here: systems biology approaches may provide a means of integrating these networks into a more complete understanding of cellular function, thus making it possible to more accurately classify a gene's functional role and effects or at least to locate its function along a continuum between strictly regulatory and solely structural.

Nevertheless, two mechanisms have been commonly observed to lead to increased gene expression among co-regulated genes. The first is knocking out the regulatory function (deregulation) to cause constitutive enzyme expression. Deregulation of gene function has been cited as a key first step in the adaptation of *Klebsiella* to xylitol, a substrate on which the wild type could barely grow (Lin 1976). It is also the major mutational step involved in invasion of a novel ecological niche, the air–broth interface, in static microcosms of initially broth-colonizing *Pseudomonas fluorescens* (Bantinaki et al. 2007; McDonald et al. 2011). Deregulation has also been cited as the mechanism behind the emergence of a "fast-switching" *E. coli* genotype capable of shifting more rapidly than its ancestor from its preferred resource, glucose, to acetate when both resources are present in the media (Friesen et al. 2004; Spencer et al. 2008). Evolved populations of *E. coli* under phosphate- or glucose-limited conditions provide further examples: mutations have been observed in genes involved in known regulatory functions such as a stress response gene (*spoT*), a starvation-specific transcription initiation factor (*rpoS*), an RNA chaperone (*hfq*), and a transcriptional repressor, *nadR* (Barrick and Lenski 2009; Maharjan et al. 2010; Wang et al. 2010; Woods et al. 2006). Comparable results have been seen with yeast populations as well (Anderson et al. 2010; Kvitek and Sherlock 2011).

The second mechanism leading to increased gene expression among coregulated genes involves gene duplications and higher-order amplifications of genetic material. Gene expression increases simply because more genes are making more product. Gene amplifications have been observed under a range of selective conditions, including nutrient limitation, antibiotic selection, and thermal stress (Bergthorsson, Andersson, and Roth 2007). For example, Gresham et al. (2008) found that 15 out of 16 clones of yeast selected under sulfate limitation for about 200 generations possessed amplified regions containing *SUL1,* a high-affinity sulfate transporter. Notably, gene duplications are quite common in many microbial populations, and they have been estimated to occur between four and eight times more often than function-improving point mutations (Bergthorsson, Andersson, and Roth 2007). They are also highly unstable and can be lost rapidly. Consequently, a point mutation that confers a larger fitness effect than the duplication itself may become fixed and the duplication lost (see, e.g., Sun et al. 2009). Alternatively, a point mutation in one copy of the duplication may eventually become fixed but the non-mutated copies lost. Either way, the importance of gene duplications would be underestimated by NGS efforts that examine only the endpoint strains rather than intermediate ones.

Stress versus Growth

Microbes possess a range of mechanisms allowing them to respond to the changing conditions they face in their lifetimes. Some of these, like the expression of the *lac* operon by *E. coli* in response to the presence of lactose in the medium, are very specific. But there are higher-level responses controlling the balance between stress response and vegetative growth as well. When conditions are stressful, such as when nutrients are limiting or the population experiences abiotic conditions outside of its normal tolerance range, microbes respond by turning on genes specifically for dealing with stress and turning off those involved in growth and general housekeeping functions. When favorable conditions return, the stress genes are deactivated and housekeeping and growth functions resume. It has been suggested that maintaining a balance between stress and growth represents a fundamental trade-off that all microbes have to contend with (Ferenci and Spira 2007; Ferenci 2005).

The laboratory conditions often used in experimental evolution are almost invariably stressful for the study organism. This is in part by design, since without the stress there would be little opportunity for adaptation and not much to study. It is also due to the way that selection experiments are usually performed (see Chapter 1). Daily transfer of a batch-culture experiment leads to regular transitions between periods of nutrient abundance soon after transfer and nutrient limitation during stationary phase. Chemostat experiments effectively do away with any period of nutrient abundance, since the population is kept in a steady state of nutrient limitation for the entire experiment.

The primary physiological effect experienced by cells during a selection experiment is therefore likely to involve activation of the stress response and the impairment of vegetative growth functions. Any mutation that allows growth to resume is likely to be favored. We might expect, then, that a primary target of adaptation in experimentally evolving microbes would be mutations in stress response genes that tilt gene expression back toward a more equitable balance between stress response and growth.

There is evidence in bacteria and yeast that a common target of mutations in experimental evolution is the transcriptional machinery controlling the balance between stress responses and growth (reviewed by Dettman et al. 2012; and see Box 5.2). For example, one of the highest levels of transcriptional control is reached by adjusting RNA polymerase activity, which can be achieved through cofactors that control the general stress response (σ^s in bacteria and the Spt-Ada-Gcn5-acetyltransferase, or SAGA, complex in yeast) or through changes to DNA supercoiling, presumably by controlling the ability of promoters to bind to DNA. Dettman et al. (2012) highlighted a number of genes associated with stress-related transcription that have been mutated in selection experiments. These include, in bacteria, σ^s itself (Conrad et al. 2009; Herring et al. 2006; Kinnersley, Holben, and Rosenzweig 2009; Maharjan et al. 2006; Wang et al. 2010; Zambrano et al. 1993), other proteins modulating stress responses (*hfq, spoT*), RNA polymerase—*rpoB, rpoC* (Conrad et al. 2009; Cooper, Rozen, and Lenski 2003; Herring et al. 2006; Maharjan et al. 2010; Minty et al. 2011; Wang et al. 2010), and genes associated with DNA topology—*topA, fis, dusB* (Crozat et al. 2010, 2005). In yeast, mutations in the SAGA-mediated stress response and other targets of eukaryotic gene expression such as chromatin remodeling, transcription, translation, and cell growth have all been identified (Anderson et al. 2010; Araya et al. 2010; Kvitek and

---

Box 5.2

How Mutations in RNA Polymerase Reconfigure the Balance between Stress and Growth in E. coli

For a better understanding of what is perhaps the most comprehensive picture to date of how mutations in the transcriptional machinery readjust the stress response–growth balance during experimental evolution, let's examine one study in more depth. Conrad et al. (2010) identified three mutants, each containing short deletions to the *rpoC* gene of RNA polymerase, whose fitness had increased by 60 percent over wild type following 25 days of selection in a minimal medium containing glycerol. These fitness increases carried over to other kinds of minimal media but not to rich media, where they were actually costly and suggested that the mutations were associated with general improvements to growth under nutrient limitation. *rpoC* codes for the β subunit of RNA polymerase and plays an essential role in transcriptional activity, and further work showed that all three mutants exhibited faster transcript elongation afforded through changes that seem to allow the cell to use promoters involved in aspects of cell growth that are usually rate-limiting in minimal media. Consistent with this result, a number of key amino acids such as cysteine, isoleucine, and pyrimidine are limiting to growth in minimal media, and at least some of the genes associated with the synthesis of these amino acids were upregulated in the mutants. This result could also explain the loss of fitness in rich media—that kind of environment supplies amino acids in abundance, and so amino acid biosynthesis is likely to be costly. It's also notable that key genes associated with σ^s, such as those involved in cell adhesion and acid resistance, were also strongly downregulated in all three mutants.

Sherlock 2011). Similar results have also been noted in phage, where many mutated sites seem to be involved in transcriptional or translational regulation (Wichman and Brown 2010).

Summary

We are still some way from being able to say with any confidence which of these alternative models is more appropriate as a means of making predictions about the genetic targets of adaptation. The gaps in our knowledge lie in our still quite rudimentary understanding of gene function and our almost complete ignorance of how genetic systems are regulated and become integrated into a physiological response. Although with broad strokes we can make a little progress in identifying at least some of the genes involved, we do not yet have a complete description of how changes at the DNA level get translated into changes in fitness. Perhaps the only thing we can confidently say at the moment is that the *cis*-regulatory model favored by developmental biologists finds very little support in microbial selection experiments. There is little evidence that changes to *cis*-regulation in the narrowest sense are solely responsible for adaptation to novel environments; and there is, as yet, no evidence to date that the mechanism responsible—reduced pleiotropy relative to what would be seen in a protein-coding gene—is at work. It does seem likely, however, that a somewhat softer interpretation involving changes to patterns of gene regulation is often associated with the early stages of adaptation. It's also possible that the stress response–growth balance is the key mechanism driving this re-patterning of how genes are regulated.

CONCLUSIONS

This chapter began with the story of Eva Markvoort and how her life was cut short because her lungs failed. Understanding how this happened, and how to prevent it happening in those living with CF now, is one of the most important challenges in CF research today.

Although many factors contribute to declining lung function in people who have CF, there is no longer any doubt that adaptive evolution of the pathogens that chronically infect the lungs is one of these factors. For a long time, microbiologists thought that the bulk of evolution that mattered for understanding pathogenic microbes had already happened. What made *Pseudomonas aeruginosa* such a problem for human health, for example, was that it *had* evolved to be an opportunistic pathogen, not that it continued to evolve when it was in a host. It's clear that this view is no longer tenable. Certain microbes may have some unique features that make them more likely than others to cause a particular infection—the large suite of virulence factors and high intrinsic resistance to antibiotics in *Pseudomonas,* for example. Still, once they initiate an infection, a whole lot else can happen. Adaptive evolution during a chronic infection clearly matters for host health.

It's hard to resist thinking that if scientists understood adaptation to the lung environment better, we might be able to do more to treat infections associated with CF. Such a research program is already under way, although its progress is somewhat limited because there are no good animal models for CF that can be used to conduct large-scale prospective experiments. So it is useful, for the time being at least, to turn to laboratory experiments such as those reviewed in this chapter to ask what they might tell us about the general properties of adaptation and then use these principles to inform our investigations of lung infections. To the extent that adaptation does happen in the lung and the CF lung is a novel environment for *Pseudomonas,* the signature of adaptive evolution will be left in DNA sequences. Examining just what those signature changes are has been the goal of this chapter.

Genomic studies of laboratory selection experiments offer an unprecedented degree of insight into the molecular details of adaptation. These details often appear overwhelming at first, especially for evolutionary biologists and ecologists who are more accustomed to thinking about issues at the population or species levels. But some compelling patterns are beginning to emerge that can help us start making sense of this data:

1. Relatively few SNPs are substituted during adaptation.

2. The rate of SNP substitution declines with time, at least over the short term of between tens and thousands of generations.

3. Epistasis at the level of fitness among substituted SNPs is fairly common and can sometimes restrict the genetic trajectories taken.

4. Parallel evolution becomes less likely in larger genomes.

5. Adaptive mutations often involve gene expression changes that dial down stress responses to allow growth and housekeeping functions to resume.

6. Many of these gene expression changes involve mutations that change the regulation of genetic systems within the cell but do not necessarily involve regulatory genes or regulatory regions of DNA sequence.

Many of these conclusions are still based on rather scant data and so need to be verified. And of course, as with most work in experimental evolution, it's important to remember that only the simplest of environmental scenarios have been studied to date: adaptation proceeding through the substitution of mutations arising de novo in what is usually a fairly homogeneous and temporally stable environment. It remains to be seen how robust these conclusions will be when subjected to more complex scenarios.

It's also imperative to remember that these conclusions bear on one particular region of parameter space: large, asexually reproducing populations in the absence of dispersal. Selection is expected to be strong relative to drift under these conditions. It is therefore not surprising that most substitutions we observe show signatures of being selected, especially considering that most of the genomes used in experimental evolution are fairly compact and lack extensive amounts of so-called junk DNA. Thus the results reviewed here likely provide a fairly accurate picture of how genomes are expected to change in evolving microbial populations in nature, but they almost certainly do not reflect what is going on in sexually reproducing organisms with smaller effective population sizes; selection is expected to be much weaker relative to drift under these conditions. A more comprehensive view of the genomic signature of adaptation will require an expansion of the range of conditions studied in laboratory experiments to include sexual reproduction and a wider range of effective population sizes.

Drawing inferences about general features of the genomics of adaptation might be especially challenging for phage, which seem to be genuinely different from both bacteria and single-celled eukaryotes. Substitutions occur faster in phage due to their high mutation rates, and their ultracompact genomes mean that different genes often share open reading frames. As a consequence, parallel evolution is more common in phage than bacteria or fungi.

We also need to know more about the dynamics of selection itself. Much of what we know comes from the comparative analysis of founder and end-point-evolved genomes, what Dettman et al. characterized as the genome-in, genome-out approach (Dettman et al. 2012). Likely much more is going on, genomically speaking, that we are missing. How important is clonal interference, for example, in governing the genomic patterns we observe in the endpoint lines? Does it lead to the substitution of mutations with distinct phenotypes, over and above the bias toward enriching for genes with large fitness effects? What about gene duplications, which are surprising by their absence from many sequencing studies? We need more detailed studies of genomes during adaptation, and not just at the endpoints, to answer these questions.

What about the prospects for using genomics to address more practical problems of an evolutionary nature? The best-case scenario will be if the sites that are observed to respond to selection in the lab are the same ones seen to be segregating in nature, because this would suggest that adaptive evolution is using similar pathways in both cases. There is good evidence from phage that this happens—between 2.3 percent and 36 percent of variable sites are shared between the lab and nature, depending on how stringently a site is categorized as being under selection or not (Wichman and Brown 2010). There are other examples in bacteria, particularly *rpoS* mutants involved in the starvation response, and in yeast, *MTK1* mutants affecting gene expression, that appear to be commonly selected in the lab and segregating in nature (see Dettman et al. 2012). The match between lab and nature in the genetic variants being selected is an issue worth pursuing in the future.

It may be possible, however, to use some of the insights we have gained about the kinds of genes targeted during adaptation to at least guide our thinking about where to look for evolutionarily relevant genomic targets. Here it would be well to focus initially on genes that affect global expression patterns, and perhaps those in particular that govern the stress–growth trade-off, at least in microbes. Ultimately, the genomic response to selection will be governed by some combination of global regulation and specific response to particular conditions. Evaluating the extent to which these two components of the response contribute to adaptation in more natural systems will be an important avenue for future research.

It's hard to read Eva Markvoort's story about her struggle with CF and the infections that go along with it and not wonder whether understanding the genomics of chronic infection actually would have helped much. After all, she needed help with so many other, more immediate problems—a lack of lung donors probably was the most important. But genomic insights are going to help eventually, even if they offer more of a medium- or long-term fix. They allow us better insight into how the pathogen causes disease, and they can help us design better therapies. Ideally, such therapies would involve targeting the evolutionarily responsive sites and either preventing them from mutating, and so causing disease progression, or mitigating the effects of those mutations once they happen. A truly predictive and practical theory of adaptive evolution must, therefore, incorporate this genomics approach in the future.

PART TWO

☙

Diversification

Τhe first species that were identifiably living, replicating cells are thought to have arisen about 3500 million years ago. Today there are something on the order of 5 to 50 million species on Earth. How did we get from one species to the many millions that coexist on Earth today?

The answer depends on whom you ask. Evolutionary ecologists and geneticists tend to focus on speciation, the process by which a single, genetically coherent population becomes two independently evolving lineages that do not, or cannot, exchange genetic material. The defining feature of diversification to an evolutionary ecologist is thus the evolution of reproductive isolation; broader patterns of diversification at higher taxonomic levels are of less concern. If we can explain speciation, the thinking goes, we can also explain patterns of diversity at higher taxonomic scales. Here, quantitative models, mostly built on an architecture provided by quantitative genetics, have taken us some way toward understanding how selection, both natural and sexual, as well as ecological interactions like competition can lead to phenotypic divergence and lineage splitting. However, they remain poor tools for asking questions above the species level, like how many species are expected to evolve, and how fast this should happen.

Paleontologists and comparative biologists, on the other hand, are less concerned with the details of speciation. Instead, they start by assuming that it happens. They then ask what governs the rate of speciation (or, more accurately, origination if the focus is on taxa higher than species) and extinction, because it's the difference between these two rates that sets the overall rate of diversification. This approach has been successful at identifying some broad features of the biology of some groups that are associated with elevated rates of diversification, such as the evolution of phytophagy in insects (Mitter, Farrell, and Wiegmann 1988) and sexual selection in some birds and fish (Ritchie 2007). But perhaps its most important role is in providing an empirical description of how diversification rates vary in space, time, or across phylogenetic groups (Benton 2009; Benton and Emerson 2007; Erwin 2009; Ezard et al. 2011; Stanley 2007). Explaining such variation is difficult, however, simply because there's usually no way of doing prospective experiments with phylogenies or fossils.

It isn't hard to see why these two approaches to understanding diversification have ended up talking past each other on many occasions: their primary goals are different. Yet ultimately they are dealing with the same underlying process, one that starts with lineage splitting within a species and then progresses to genuine speciation. Repeating this process many times gives us the branching pattern we see in phylogenetic trees. The mechanisms governing how diversification leads to the tree of life and why some branches of that tree become so much more diverse than others is the major goal of diversification studies. It is also the focus of this part of the book.

The aim here is to begin outlining a more empirically grounded theory of ecological diversification. In Part One, Adaptation, the basic elements and building blocks of the theory were put in place. There we saw that divergent selection toward different phenotypic optima constituted the beginning step in adaptive diversification because it generates niche specialist types. This second part of the book takes niche specialization as the starting point and asks how effective this process, which involves lineage splitting from one to two (or at most a few) ecologically distinct populations, is at generating some of the more familiar patterns of lineage diversification seen in comparative and paleontological studies. At the same time, this discussion takes advantage of the availability of an increasing number of genomic studies of diversification to ask whether any general rules govern what and how many genes are involved in diversification.

The book draws on the microbial selection experiment literature for examples and for data with which to test the key predictions. Experimental evolution affords a unique opportunity to study the process of diversification in a defined ecological setting. By contrast, the inability of many comparative studies to characterize adequately the ecological conditions under which diversification occurs is a major limitation of this approach and is one of the main reasons for the continued rift between microevolution and macroevolution. By defining the ecological conditions more precisely at the outset and then following the emergence and fate of diversity through time with fast-evolving microbial populations, we can effectively do the sort of experiment that Stephen Jay Gould thought could never be done: rewind the tape of life and play it again.

Of course, this does not mean that microbial experiments have all the answers. The usual caveats still apply. Most worrisome for many researchers not familiar with microbial experimental evolution is the distinct absence of speciation from most microbial diversification studies. The reason for this is simple: the experiments have, for the most part, not been done. The bulk of microbial experiments on diversification to date have been conducted with bacteria, which are asexual haploids. Formally, the theory for the evolution of ecological diversity does not require a distinction between modes of reproduction to be made. Still, much research has focused on speciation as a limiting step in the process of diversification because it's generally viewed as the harder process, next to ecological differentiation, to understand. The truth of this statement, as well as the importance of ecological differentiation to speciation itself, remains an empirical question.

The experiments reviewed here take us some way toward gaining a clearer picture of how evolutionary diversification occurs. These experimental systems therefore provide a bridge between the more traditional approaches to the study of diversity at the level of populations or incipient species favored by evolutionary ecologists and geneticists and the broader-scale descriptions of diversity associated with macroevolutionary approaches. The work reviewed here offers a unique glimpse into how ecological diversity arises, as well as its fate. Through this process, the hope is that the microbial experimental evolution research program can generate new hypotheses that can be used to improve our understanding of the general principles governing evolutionary diversification.

CHAPTER 6

Phenotypic Disparity

The sun is setting on the edge of a grassy plain. The wind is dry and warming. Gently rolling hills of gold, green, and ocher stretch to the horizon. The grasses, about the height of a man's chest, are so rich in species that a naturalist would be occupied for years trying to identify them all. This is a place of incredible natural beauty and variety of life. It is also, quite possibly, the birthplace of humankind.

The site is Maropeng, about 40 kilometers from Johannesburg, South Africa. In 2010, Lee Berger, a professor of paleoanthropology at the University of Witwatersrand, and his research team announced the discovery of four partial skeletons from a fossil site in a limestone cave. What the researchers have been learning about these fossils, which have been named *Australopithecus sediba,* or wellspring southern ape, is changing what we know about how modern humans evolved.

Paleontologists have long thought that modern humans, *Homo sapiens,* evolved from one of the five extinct apelike human species collectively known as *Australopithecus.* Both *Homo* and *Australopithecus* walked on two feet, but *Australopithecus* had much smaller skulls than *Homo.* The difference in skull size was about the same as that between an apple and a grapefruit. For years, paleontologists have been trying to piece together the puzzle of which particular *Australopithecus* species underwent the shift from small brains to the much larger brains of *Homo.*

With Berger's discovery, the puzzle has become a lot more complicated. *A. sediba* is a sixth species that dates from approximately 2 million years ago, about the time paleontologists think *Australopithecus* was in the process of becoming *Homo.* The skeletons that have been described in most detail—one appears to be a 9- to 13-year-old boy and the other an adult female—have skulls with brains about the size of a typical *Australopithecus,* but their jawbones and teeth resemble those of *Homo.* And there are other oddities as well.

The specimens have human-like ankles, but apelike heels. Their diet, which the researchers investigated by removing small bits of plants that were preserved in the hardened plaque covering the teeth, included typically human foods like fruits but also more apelike foods like tree leaves, wood, bark, and grasses. Like the mythological chimera, that combination of lion, serpent, and goat, *A. sediba* has traits it seems to have borrowed from both *Australopithecus* and *Homo.* This makes for no small amount of confusion among paleontologists.

The confusion stems partly from disagreements over what the new fossils should be called. The weird mix of traits in *A. sediba* makes them difficult to classify. Are they genuinely a new species of *Australopithecus,* or are they actually an early species of *Homo?* Some paleontologists, like Donald Johansen of Arizona State University, feel that *A. sediba* is human enough to warrant being *Homo.* Others, like John Hawks from the University of Wisconsin–Madison, feel that the name *Australopithecus* is appropriate. "A new species within *Australopithecus* was probably the right call, but not an easy one," he writes on his blog (john hawks weblog).

Nomenclature aside, there is a deeper issue here about human origins. It's widely appreciated that the major transition between *Australopithecus* and *Homo* involved a switch from a primarily forest-dwelling life to a more terrestrial life. *A. sediba,* with its chimeric qualities of ape and human, clearly represents a transitional form between these two ecological niches. Still, the causes of this transition remain unclear. The transition may have happened because of changes in the physical environment—the creation of more extensive savanna and grasslands thanks to continental drift, for example—or because of competition among the early human species coexisting in the same environment.

Although paleontologists have traditionally held that the environment was the major driver of human evolution, the discovery of *A. sediba* makes it clear that competition cannot be discounted. *A. sediba* lived at the same time and in the same place as other *Australopithecus* species and would have interacted and competed with them for resources like food and shelter. It's possible that *A. sediba's* peculiar mix of traits evolved because those other species were present and competing for similar resources. If so, then this raises the possibility that who we are today is largely the result of who our neighbors were 2 million years ago.

This is not a new idea, of course. Evolutionary ecologists have long known that competition can play a major role in phenotypic divergence. Perhaps the best evidence comes from the canonical examples of adaptive radiation in sticklebacks, Galápagos finches, and Caribbean lizards where divergence in traits related to resource use is often tied to speciation. The basic idea here is that competition is an agent of selection: individuals with similar ecological phenotypes—traits like beak size or body depth, associated with capturing and using resources—compete most intensely for resources. Selection will then favor dissimilar phenotypes that use alternative resources, reducing the strength of competition between them in the process.

It's tempting to think that competition might have played such a role in the diversification of early humans. The idea remains controversial, though, because what we know about our early human ancestors is limited to only a handful of fossils. Berger's discovery stands to change this view by providing much more data. But no matter how much fossil data we have, there's always a limit to how much we can infer. Fossils can tell us what happened in evolutionary history, but not how it happened. For that, we still need experiments.

This chapter reviews what we know from microbial selection experiments about the various mechanisms driving ecological and phenotypic diversification. Is divergent selection driven by different environmental conditions, like the forest versus the savanna, strong enough to cause substantial phenotypic differentiation like that seen between *Australopithecus* and *Homo?* Or do we also need to invoke competition as an additional source of selection, as the preliminary data from *A. sediba* suggests? Microbial experiments obviously cannot tell us anything about the details of early human evolution, but they can help us evaluate the plausibility and generality of these hypotheses. We're more likely to believe that competition drove diversification of early humans if multiple lines of evidence—fossil data, comparative data from other species, and experimental data from microbes—are all pointing in the same direction.

THE EVOLUTION OF PHENOTYPIC DISPARITY

Those who study comparative morphology commonly use the term *phenotypic disparity* to mean "a measure of phenotypic divergence or variation among individuals or lineages." The most straightforward measure of disparity is either a scaled difference, for example in terms of numbers of standard deviations from the mean, or a variance. Paleontologists and comparative biologists use phenotypic disparity to express how different the traits of fossil or contemporary species are from each other or their supposed common ancestor. In comparative studies of adaptive radiation, the traits being measured are often implicitly assumed to correlate in some way to fitness in different environments, and so to reflect the degree of ecological specialization. In microbial experiments, it is easier to obtain direct measures of fitness so measures of phenotypic disparity can be made directly in relation to fitness variation. The following discussion assumes that disparity refers to fitness itself, or traits closely related to fitness.

Four factors contribute to the evolution of phenotypic disparity: genetic drift, genetic interactions that create rugged fitness landscapes, divergent selection, and interactions among individuals involving access to resources or mates. The effects of these factors are cumulative, so phenotypic disparity should increase through time. **Figure 6.1** illustrates how each factor can contribute to the evolution of phenotypic disparity. These are phenotype-through-time plots similar to those used by paleontologists (see, e.g., Erwin

2007; Foote 1993). They illustrate the course of phenotypic evolution from a common ancestor through time. A plot is interpreted as follows. The phenotype of the common ancestor to a group is located at an arbitrary position in phenotype space at the bottom of each panel. The lines connecting the ancestor to daughter lineages (or species) represent the phenotypic pathways taken by a particular lineage over the course of its evolution from the common ancestor. For illustrative purposes, we'll follow the evolutionary trajectory of a handful of independently evolving lineages derived from the same common ancestor. In effect, we are rewinding the tape of life, as Stephen Jay Gould was fond of saying, and playing it again. In this figure, the phenotypic spaces occupied by extant populations are represented by the solid shapes across the top of each panel.

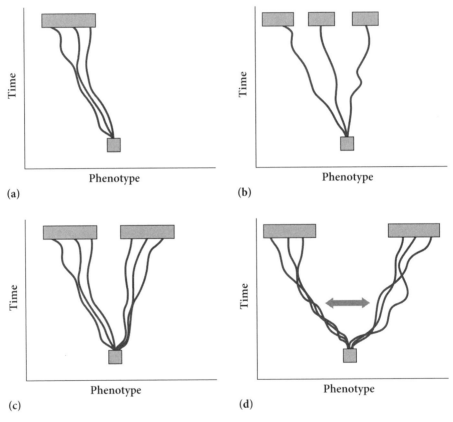

Figure 6.1 The components of diversification. Black lines are possible evolutionary routes to distinct phenotypes that connect the ancestor (the small gray square at the base of each plot) with descendant lineages (the gray boxes across the top of each panel). The size and position of these boxes indicate the extent of phenotypic disparity. Divergence can occur due to (a) stochastic processes alone; (b) sign epistasis, which can generate gaps in the phenotype space; (c) divergent selection, leading to two distinctly adapted niche specialists; (d) disruptive selection being driven by ecological interactions (the gray bidirectional arrow).

To understand how each of the four factors just mentioned contributes to the evolution of phenotypic disparity, let's imagine that we are following the phenotypic evolution of an initially genetically and phenotypically homogeneous founding population that for some reason finds itself in a novel environment to which it is poorly adapted. Under these conditions, selection is likely to be strong enough to cause a deterministic change in phenotype, as indicated by the fact that the mean phenotypes of the extant populations in Figure 6.1a are to the left of the mean phenotype of the ancestor. Nevertheless, because population sizes are inevitably finite, genetic drift associated with stochastic changes in allele frequencies will cause replicate populations to diverge from each other through time (Figure 6.1a). Keep in mind that the effects of drift on phenotypic divergence will be more pronounced in small populations than in large ones. An additional stochastic process, the chance occurrence of mutations, contributes further to divergence among populations when adaptation occurs through mutations.

Genetic interactions among mutations, or epistasis, can also contribute to phenotypic disparity by creating gaps in the range of viable phenotypes produced (Figure 6.1b). These gaps represent combinations of genes that have low fitness and so are selectively removed from the population when they occur. For this to happen, epistasis must cause the sign of the selection coefficient associated with a mutation to change depending on the genetic background in which that mutation occurs. This is called sign epistasis, and it generates a rugged underlying adaptive landscape with many peaks and valleys. The ruggedness of the landscape reflects the degree to which some genetic combinations "work" together and others do not.

Although genetic drift and sign epistasis can cause the phenotypes of replicate lineages to diverge in a uniform environment, the contribution of these processes to the phenotypic distinctiveness among species is usually thought to be rather modest. Instead, divergent selection, often caused by the presence of different phenotypic optima in the environment in the form of distinct niches or resources, is thought to be a major driver of phenotypic disparity (Figure 6.1c). In Chapters 2 and 3, we considered the effects of divergent selection on ecological specialization and local adaptation in detail. Recall that when divergent selection is strong relative to gene flow, niche specialists readily evolve. As a general rule, we expect that the extent of disparity should evolve to match the differences in the location of the phenotypic optima that are the cause of divergent selection. Divergent selection is therefore a potentially effective means of generating phenotypic disparity.

Finally, it is often suggested that some kinds of interactions among individuals can promote phenotypic divergence, and especially ecological specialization, because they cause intermediate phenotypes to suffer disproportionately relative to more extreme phenotypes (Day and Young 2004; Doebeli and Dieckmann 2000). The sorts of interactions that are usually discussed in this context are often ecological, involving resource competition, predation, or even mutualism. Sexual selection, the competition for mates, can also promote ecological diversification either by helping to reduce gene flow among habitats (e.g., through assortative mating) or by speeding the rate of adaptation to novel environments (Ritchie 2007). When underlying divergent selection is already promoting ecological specialization, these sorts of interactions can both accelerate the rate of phenotypic divergence and lead to exaggerated phenotypic divergence. Both ecological interactions and sexual selection can therefore provide an extra "push" in addition to divergent selection that can further drive populations apart in phenotypic space (Figure 6.1d).

The importance attributed to each of these processes as causes of phenotypic divergence has varied over the years. Stephen Jay Gould thought that chance was an important determinant of which metazoan lineages survived the Cambrian Explosion (Gould 1989). Ernst Mayr similarly thought that drift could play an important role in setting up the conditions to drive speciation (Mayr 1963). Others downplayed the role of chance events. Wright's use of the adaptive landscape metaphor famously put epistasis at center stage in diversification (Wright 1982), and ecologists have greatly emphasized the role of habitat heterogeneity and competition as engines of diversification (Levins 1968; Rosenzweig 1995). The task for empiricists and experimentalists is to understand how, for any given system, these different processes interact to generate phenotypic divergence.

TESTING THE THEORY

Data from microbial selection experiments can be used to test the predictions of the theory just described. The structure of replication within and among treatments makes it possible to infer the contributions of the different factors involved in generating phenotypic disparity. Variation in fitness or phenotype among replicate, independently evolving lineages within a treatment allows investigators to estimate the degree to which stochastic processes, and perhaps also epistasis, contribute to phenotypic divergence. The treatments themselves offer information on the impact of divergent selection, ecological interactions, or sexual selection. By comparing the magnitude of the variation contributed by the treatment effects to that introduced through stochastic processes (represented as the variance among replicate lineages), we can get an idea of how effective different processes are in generating phenotypic divergence. The dynamics of these different variances provides information on the rates at which phenotypic disparity evolves.

Thus a rich body of information on the evolution of phenotypic disparity can be mined from these experiments. This section summarizes some of the more compelling results gathered to date.

Phenotypic Divergence among Replicate Populations

When selection is mutation limited, as it almost always is in microbial selection experiments, both mutation and genetic drift will cause replicate populations to diverge from each other because, at base, they are stochastic processes. No two replicate populations are likely to substitute the same mutation at exactly the same time. The timing and order with which mutations occur and are substituted therefore causes replicate populations to diverge phenotypically over time. Drift also plays a role by causing beneficial mutations to be lost soon after they occur, when they are rare, and by causing fluctuations in the frequency of alleles during substitution, especially in small populations. Because the effects of mutation and drift are cumulative, replicate populations will become increasingly divergent over time. Moreover, the effects of stochastic processes are more noticeable in small populations compared to large, so there should be a negative relationship between divergence and effective population size.

To test these two predictions, data was collected on phenotypic divergence among replicate selection lines in microbial experiments. The relevant studies are shown in **Table 6.1**.

Let's consider divergence in fitness first. **Figure 6.2** shows how the variance in fitness changes as a function of time and effective population size

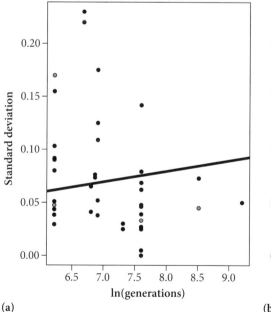

(a)

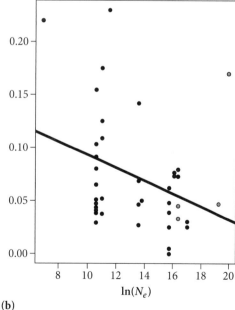

(b)

Figure 6.2 Fitness divergence among replicate selection lines as a function of the natural logarithm (ln) of time, measured in terms of generations, (a) and effective population size, *Ne* (b). Fitness divergence is measured as the standard deviation in fitness among replicate selection lines. Divergence increases with time and decreases with effective population size. Black points are haploids; gray points are diploids.

Table 6.1 Studies Reporting Fitness Variation among Replicate Selection Lines

Citation	Species	Environment	Increase in fitness (w)	Standard deviation among replicates	Generations	Effective population size (N_e)	Number of replicates
BACTERIA							
Lenski and Travisano 1994	*E. coli*	glucose	0.48	0.05	10000	990099	12
Bennett, Lenski, and Mittler 1992	*E. coli*	32°C	0.10	0.07	2000	792079	6
		37°C	0.08	0.05			
		42°C	0.19	0.14			
Cooper and Lenski 2010	*E. coli*	glucose	0.21	0.03	2000	6930693	6
		maltose	0.27	0.00			5
		lactose	0.42	0.03			6
Melnyk and Kassen 2011	*P. fluorescens*	glucose	0.12	0.03	500	39604	4
			0.14	0.04			
			0.13	0.09			
			0.09	0.05			
			0.11	0.04			
			0.18	0.04	900		
		xylose	0.27	0.09	500	39604	4
			0.38	0.15			
			0.21	0.10			
			0.17	0.05			
			0.32	0.08			
			0.35	0.07	900		
Bailey and Kassen 2012	*P. fluorescens*	glucose	0.18	0.04	1000	59406	3

Citation	Species	Environment	Increase in fitness (w)	Standard deviation among replicates	Generations	Effective population size (N_e)	Number of replicates
		malrose	0.20	0.11			
		xylose	0.22	0.18			
Baltrus, Guillemin, and Philips 2008	*Helicobacter pylori*	*brucella* broth	0.62 / 0.72	0.07 / 0.08	960	30000000	9
FUNGI							
Schoustra et al. 2009	*Aspergillus nidulans*	complete medium	0.48 / 0.38	0.23 / 0.22	800 / 800	99950 / 1000	58 / 60
Zeyl 2005	*Saccharomyces cerevisiae*	dextrose	0.81 / 0.57	0.08 / 0.03	2000	13000000	5
Zeyl 2003	*Saccharomyces cerevisiae*	dextrose	0.86 / 0.81	0.07 / 0.05	5000	13000000	5
Dettman et al. 2007	*Saccharomyces cerevisiae*	salt / glucose	0.44 / 0.13	0.17 / 0.05	500	463000000 / 234000000	6
Dettman, Anderson, and Kohn 2008	*Neurospora crassa*	salt / low temperature (12°C)	0.04 / 0.05	0.03 / 0.03	1500	25000000 / 25000000	3

used in the experiment. Variance in fitness increases with time (Figure 6.2a: *slope* ± se = 0.010 ± 0.001; $P < .0001$; note that a regression is calculated here that's fixed at the origin, because all replicate lines are descended from a common ancestor and so by definition do not vary at the start of the experiment) and decreases with increasing effective population size (Figure 6.2b: *slope* + se = −0.006 + 0.003; $P = .0304$), as expected. The stochastic processes of mutation and genetic drift thus contribute in predictable ways to the divergence among replicate selection lines.

Fitness is a conservative estimate of the extent of phenotypic disparity, because a population may include many different phenotypes with similar fitness. Life-history trade-offs, such as between fecundity and longevity, are classic examples: similar fitness can come about by having many offspring but dying young or by having fewer offspring and living longer. So it's worth asking whether similar patterns to those seen for fitness emerge when we examine non-fitness traits in microbial experiments, if only because these are the sorts of traits most commonly examined in nonmicrobial studies.

The answer to that question is yes, although admittedly a tentative yes, because so far only a handful of experiments have addressed the evolution of non-fitness traits in microbes (Melnyk and Kassen 2011; Travisano et al. 1995). Lenski and Travisano (1994) found that variance in cell volume among replicate lines of evolving *Escherichia coli* increased over time, just as it did for fitness. They also noted that the among-line variance for cell volume was about 20 percent greater than it was for fitness. This sort of pattern—higher variance for traits more distantly related to fitness—fits well with the idea that genetic and phenotypic variances should be larger for traits under weaker selection (Fisher 1930; Mousseau and Roff 1987).

More recently, using techniques for analyzing multivariate phenotypic divergence, Schoustra et al. (2012) examined phenotypic divergence in four non-fitness traits after adapting many replicate lines of the filamentous fungus *Aspergillus nidulans* to a common environment over 800 generations. They found that small populations did tend to be more divergent than large populations, as expected, but this difference was not statistically significant. Divergence proceeded along many different pathways with no obvious constraints on what kinds of phenotypes evolved. This interpretation suggests that the pleiotropic effects of beneficial substitutions can be quite variable, causing divergence in many different phenotypic directions as a correlated response to improvements in fitness. It's notable, moreover, that the lack of constraints on phenotypes in this experiment contrasts with results in populations from more complex organisms, where selection acts on standing genetic variance, where phenotypic divergence is more constrained (Walsh and Blows 2009). Thus, compared to adaptation from standing variation, mutation-driven adaptation may be much freer in terms of phenotypes that can be explored.

Epistasis

Epistasis occurs whenever the phenotypic effect of two (or more) mutations together is different from the sum of their independent effects. This very general definition applies equally to both fitness and non-fitness traits. Here we'll focus exclusively on fitness.

Epistasis for fitness can take a range of forms. If the combined effect of two mutations is greater than the sum of their independent effects, we say that epistasis is positive. The reverse situation, where the combined effect of two mutations is less than the sum of their effects on their own, is called negative epistasis. An extreme form of negative epistasis, called sign epistasis, takes place when mutations that increase fitness on their own actually decrease fitness when they occur together. Sign epistasis is especially important for understanding phenotypic divergence, because it can act as a diversifying force directing an evolving population down certain genetic pathways and not others. Sign epistasis can therefore dramatically change the evolutionary route taken by independently evolving lineages.

Two microbial experiments illustrate the influence of sign epistasis on phenotypic divergence. Salverda et al. (2011) showed that multistep resistance to cefotaxime and related drugs in *E. coli* usually involves a first mutation occurring at amino acid position 238 of the enzyme TEM beta-lactamase. Occasionally, though, a lineage evolved resistance through an alternative route whose first genetic change involved a mutation at amino acid 164 of the same gene. In combination, these two mutations actually made the cells *less* resistant to the drug and so were very unlikely ever to be found in the same genome. Moreover, which mutation occurred first mattered for subsequent evolution: those lines that substituted the 238 mutation first were about 1.5 times more resistant to cefotaxime than those that substituted the 164 mutation first.

In yeast, we see a similar phenomenon between two different mutations involved in transporting glucose into the cell, although this time the mutations are in different genes. One mutation is a non-synonymous substitution in the *mth* gene, a negative regulator of a glucose-sensing pathway, and the other involves a duplication of *HXT6/7*, a glucose transporter. Both mutations on their own increase fitness in a glucose-limited environment, but when they occur together, they actually decrease fitness. It's not clear why this happens, but one hypothesis is that the cell is just wasting too much energy because both mutations result in the overproduction of a glucose transporter.

These examples illustrate that sign epistasis occurs in microbial experiments, but we will need many more examples before we can say how common it is. The observation of marked phenotypic divergence within and among populations evolving in what appear to be fairly homogeneous environmental conditions could be an indication that sign epistasis is actually quite common.

Two such examples are the metabolic differentiation that occurs in glucose-limited chemostats of *E. coli* (Maharjan et al. 2006; Notley-McRobb and Ferenci 1999b) and in batch cultures of *Pseudomonas fluorescens* evolving on xylose as a sole carbon source (Melnyk and Kassen 2011). It's hard to interpret the results of these experiments, both involving phenotypic diversification in what otherwise appears to be a homogeneous, uniform environment, as being due to anything but reciprocal sign epistasis. Admittedly, though, it's always possible that some other, as yet unidentified, agent of selection or cryptic niche differentiation is responsible. Until more studies are available, we will have to content ourselves with the knowledge that sign epistasis can, in two instances at least, direct a population on very distinct genetic and, presumably, phenotypic pathways.

Divergent Selection

Divergent selection drives phenotypic divergence to an extent governed by the difference in phenotypic optima: the more different the optima, the more distinct the phenotypes. In Chapter 2, we discussed the effect of divergent selection on adaptation and the evolution of specialization. In short, when gene flow is absent, divergent selection is exceedingly effective at generating specialization to alternative environments in the laboratory. An example of this process, using real data from a recently published experiment (Bailey and Kassen 2012), is shown as a phenotype-through-time plot in **Figure 6.3**.

When gene flow is present, it should reduce the effectiveness of divergent selection at generating specialization. Maladapted genotypes are introduced into a patch by in-migration from elsewhere while locally adapted genotypes generated by selection are removed by out-migration. Few experiments have been done to test this prediction. Cuevas, Moya, and Elena (2003) found that migration reduced the extent of adaptation and niche specialization to host-cell types in experimentally evolved populations of vesicular stomatitis viruses (VSV), and Habets et al. (2006) showed that gene flow among spatially separated populations on the surface of agar plates prevented diversification in populations of *E. coli*. Moreover, work with *P. fluorescens* specialization to alternative resources in a spatially fine-grained environment can be prevented when one resource is much more productive than the other, because adaptation preferentially occurs in the more productive patch (Jasmin and Kassen 2007a). These results support the idea that gene flow plays a predominantly homogenizing force, preventing diversification.

Still, gene flow is not exclusively a force that prevents diversification. If divergent selection is strong enough, genetic and phenotypic divergence can occur in the face of gene flow. We saw in Table 4.3, for example, some experiments demonstrating that diversity within a population is often greater in

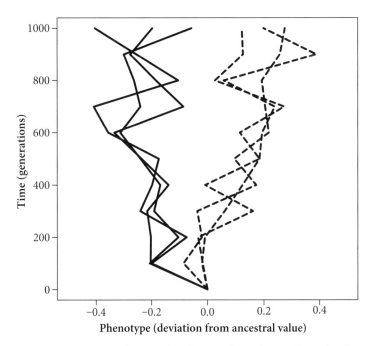

Figure 6.3 A phenotype-through-time plot showing how three independently evolving replicate populations diverge in phenotype from a common ancestor as a result of selection on xylose (solid lines) or mannose (dashed lines). The phenotype measured is fitness, and the magnitude of change is expressed relative to the ancestral value of 0. The sign of the phenotype is arbitrary.

spatially variable environments with gene flow compared to comparable uniform environments. The heterogeneity in conditions of growth across patches in these experiments thus generates divergent selection strong enough to overcome the homogenizing effects of gene flow. A small amount of gene flow may also be important in allowing adaptation and differentiation to marginal habitats where, in the absence of migrants, population sizes are so low that the supply of beneficial mutations is effectively zero. In these situations, migration from more productive habitats can introduce mutations that are occasionally beneficial in the marginal habitat and so effectively rescue the population from extinction and promote adaptation. Examples of this effect have been studied by Perron, Gonzalez, and Buckling (2007) as well as Venail et al. (2008).

It has also been suggested that strong divergent selection can be generated when a population must adapt to multiple stressors or niche dimensions simultaneously. In such situations, selection is said to be multifarious—it acts on more than one niche dimension or trait at once (Nosil and Harmon 2009; Nosil and Sandoval 2008). Phenotypic differentiation might be expected to evolve more readily when selection is multifarious, though few if any experimental tests have addressed this idea in microbial systems.

Strong, persistent divergent selection that drives phenotypic differentiation can cause speciation if reproductive isolation tends to evolve as a by-product of adaptation to different environments. This process, known as ecological speciation, is thought to be the major factor in the evolution of reproductive isolation (Rundle and Nosil 2005; Schluter 2000a). Two studies of facultatively sexual fungi, one in yeast (Dettman et al. 2007) and the other in *Neurospora* (Dettman, Anderson, and Kohn 2008), have documented experimental results consistent with ecological speciation. In both studies, a few hundred generations of divergent selection in the asexual phase were sufficient to generate ecological specialization and, as a consequence, postmating isolation. It's important to note that the evidence for reproductive isolation came from comparing the fitness of hybrid offspring between pairs of evolved lines that had evolved either in the same environment or in different environments. Hybrids from different environments had substantially lower fitness than those from the same environment, providing unequivocal evidence for postmating reproductive isolation as the result of ecological specialization.

Further work with the evolved populations of yeast showed that the cause of reduced hybrid fitness was what's known as a Dobzhansky-Muller (DM) incompatibility (Anderson et al. 2010). DM incompatibilities arise because adaptation to different environments results in the fixation of mutations at different loci in each environment. When these mutations are brought together in a hybrid, their combined effects—having never been tested by selection—result in a decrease in fitness. The alleles responsible for the DM incompatibility in the yeast experiment were a proton efflux pump from a highly salt-adapted line (*PMA1*) and a global regulator of mRNA production from a low-glucose line (*MTK1*). However, the physiological cause of reduced fitness remains a mystery.

Ecological Interactions

As we have seen, it's relatively easy for gene flow to prevent diversification unless divergent selection is very strong. For this reason, it is often thought that some additional force, acting on its own or in concert with divergent selection imposed by the presence of different niches in the environment, is required to drive the evolution of phenotypic disparity and, eventually, speciation. One of the most attractive suggestions, partly because it's so general, is that ecological interactions such as competition and predation, and perhaps even mutualism, are responsible for providing the extra push that drives ecological and phenotypic differentiation.

The basic idea is that the interactions normally occurring among genotypes or species in a community can themselves generate strong selection that affects not only their own evolutionary trajectory but also that of other types in the community. In other words, ecological interactions generate forms of coevolution that can cause phenotypic divergence. In the generation of ecological diversity, it's crucial for selection to be disruptive: ecologically intermediate phenotypes are selected against, and extreme phenotypes are favored.

RESOURCE COMPETITION Resource competition is the most commonly cited mechanism for disruptive selection. Undoubtedly, it has played a role in a number of the hallmark cases of adaptive radiation (Schluter 2000a). Many microbial selection experiments likewise attribute the emergence of resource specialists and diversity directly to resource competition (Barrett and Bell 2006; Barrett, MacLean, and Bell 2005; Brockhurst et al. 2006; Friesen et al. 2004; Habets et al. 2006; Jasmin and Kassen 2007a). The most compelling example is the rapid and highly repeatable radiation of *P. fluorescens* into morphologically distinct niche specialist types, as documented by Rainey and Travisano (1998) and discussed in Chapter 1. Recall that diversity among the three main types—the broth-colonizing *smooth,* the biofilm-forming *wrinkly-spreader,* and *fuzzy-spreaders* that form rafts at the air–broth interface but then fall and accumulate in the anoxic zone at the bottom of the microcosm—is stably maintained by negative frequency-dependent selection. Rare *smooths* can invade a population of *wrinkly-spreaders,* for example, and vice versa (the same is true for a *smooth* and a *fuzzy-spreader* although, interestingly, not between a *wrinkly-spreader* and a *fuzzy-spreader*).

Resource competition is often implicated as the mechanism driving diversification in this system, and in all the other examples mentioned, simply because alternative mechanisms such as predation or parasitism are excluded by design. A more direct test would involve manipulating the strength of competition itself and then following the evolution of diversity. The prediction would be that stronger competition leads to more pronounced phenotypic and ecological divergence. Currently, no experiment has done this. The closest might be a recent experiment in which strains of *P. fluorescens* carrying different antibiotic resistance mutations were allowed to diversify in spatially structured microcosms (Bailey et al. 2013). The resistance mutations reduced the density of the diversifying population, but otherwise had no other detectable effects. Low-density strains produced fewer derived colony morphotypes than high-density ones, consistent with the idea that resource competition is weaker when densities are lower. Note, however, that the endpoint measure here is the number of morphologically distinct types, not phenotypic disparity. Thus the results should be regarded with caution for what they say about the effect of resource competition on phenotypic divergence in the strictest sense.

Disruptive selection against intermediate phenotypes is the mechanism by which phenotypic divergence should happen, although this process has never been directly demonstrated experimentally in a microbial system. The best data available comes from work by Tyerman et al. (2008), who showed that character displacement—more extreme phenotypic divergence among individuals in sympatry than in allopatry—evolved when populations of *E. coli* are propagated on a mixture of glucose and lactose over a thousand generations. Tyerman et al. found that the growth profiles of two distinct genotypes became more similar when allowed to evolve in isolation compared to when they were propagated together. This result can be explained only if resource competition was a driver of phenotypic diversification.

SHARED ENEMIES   Shared enemies such as predators or parasites have long been suggested to be important causes of phenotypic divergence in prey or hosts (Ehrlich and Raven 1964; Vermeij 1994). Enemies cause selection for resistance to attack in their victims, which in turn generates selection on the enemies to find new ways of attacking victims. Enemy and victim effectively act as constantly changing sources of divergent selection for each other, a process known as antagonistic coevolution. Phenotypic divergence in prey can also happen if enemies act as a source of disruptive selection, because intermediate phenotypes are more susceptible to predation or parasitism than extreme phenotypes are (Vamosi 2005).

Microbial selection experiments provide many examples of shared enemies driving diversification in victim populations (Bohannan and Lenski 2000; Brockhurst, Buckling, and Rainey 2005; Brockhurst, Rainey, and Buckling 2004; Buckling and Rainey 2002b; Chao, Levin, and Stewart 1977; Gallet et al. 2007; Lenski and Levin 1985; Schrag and Mittler 1996). Victim diversification occurs due to a trade-off between resistance to the enemy and a cost of resistance in the absence of enemies, leading often to coexistence between susceptible and resistant prey. Most of these studies involve bacteriophage infections of bacteria, and so constitute models of parasite–host interactions. A few studies have used a genuine predator, the protozoan *Tetrahymena thermophila*, and bacterial prey (Friman et al. 2008; Hall, Meyer, and Kassen 2008; Meyer and Kassen 2007) and have found similar results.

The best example to date of antagonistic coevolution comes from Buckling and Rainey (2002a), who demonstrated repeated cycles of antagonistic coevolution between the bacterial host, *P. fluorescens,* and a lytic phage, φ2, over 300 to 400 bacterial generations. Both host and parasite became more generalized in their interactions: evolved bacteria were resistant to a broader range of phage, and evolved phage could infect a broader range of bacteria. Antagonistic coevolution was thus underlain by changes in a wide range of genomic targets in both species, a result later confirmed by whole genome sequencing (Paterson et al. 2010). Despite the trend to infect and be infected by more types, coevolved bacteria and phage did show some specificity in their interactions: bacteria were better able to resist phage from their own microcosm than from different microcosms. Thus coevolution led to a degree of evolutionary divergence among replicate populations. The genetic causes of this divergence, whether due to stochastic processes such as mutation order or due to specific interactions between host and parasite (a form of genotype-by-genotype interaction), have not been explored.

Repeated cycles of antagonistic coevolution are not commonly observed in *E. coli* studies (but see Marston et al. 2012; Meyer et al. 2012; Schulte et al. 2010), although they have been seen in other systems (Bérénos, Wegner, and Schmid-Hempel 2011; Lohse, Gutierrez, and Kaltz 2006). The reason why cycles are not seen in *E. coli* experiments is not clear. Phage infecting *E. coli* often cannot overcome bacterial resistance even though their population sizes are larger than those of bacteria. One explanation is that the small size of phage genomes imposes a constraint that limits the evolutionary options available to phage. This interpretation is easily testable by asking whether phage with larger genomes tend to be more likely to coevolve with *E. coli*.

Antagonistic coevolution has also not been extensively studied in genuine predator-prey systems. Only one experiment has been conducted for long enough to observe antagonistic coevolution, but in this case the predator, the protozoan *Tetrahymena thermophila*, showed no evidence of having evolved even after 2400 generations of evolution in the prey, *Serratia marcescens* (Friman et al. 2008). Predator population sizes in this system are much smaller than prey, and generation times are substantially longer; these two factors together may have contributed to a lack of evolutionary response.

Currently, no microbial selection experiments have demonstrated disruptive selection by shared enemies as a cause of diversification in victim populations. Even so, some evidence from field studies in *Daphnia* indicates that this can occur (Duffy et al. 2008). It's possible that the lack of good experimental examples of disruptive selection by predators is something of an artifact of studying systems where resistance to predation is fairly discrete because it is the result of one or at most a few mutations. If so, it will be worth expanding the range of model systems for studying the causes of phenotypic divergence in prey to include those with a more complicated underlying genetic basis for resistance.

MUTUALISM   Some researchers have suggested that mutualism can generate disruptive selection within one or both of the interacting species (Doebeli and Dieckmann 2000). Diversification arises due to a trade-off between being able to exploit one's own resources and benefiting from the resources supplied by the mutualist. Two recent studies have demonstrated the de novo evolution of mutualisms between distinct microbial species (Harcombe 2010; Hillesland and Stahl 2010), but no known experiments have documented mutualism as a source of disruptive selection leading to phenotypic divergence.

SEXUAL SELECTION

Sexual selection occurs when variation in reproductive success is caused by competition for mates. A mutation that increases mating success, either because it allows the bearer to mate more often or have more successful fertilizations, will spread through a population. Biologists have often suggested that sexual selection is sufficient on its own to generate speciation. In support of this idea, comparative evidence from birds (Barraclough, Harvey, and Nee 1995), insects (Arnqvist et al. 2000), and fish (Mank 2007), shows higher rates of diversification in clades that experience stronger sexual selection. However, in comparative studies it's often difficult to disentangle sexual selection per se from ecological sources of selection, so it is unclear to what extent which process is driving diversification. Clearly, there is room here for some experiments.

In testing these ideas, it's important to be mindful of the complex ways in which sexual selection and natural selection can interact. One view sees them as antagonistic: at equilibrium, energy invested in sexual fitness should trade off against energy invested in nonsexual fitness. For example, genotypes of the unicellular alga *Chlamydomonas reinhardtii,* when selected for improved mating efficiency, had lower intrinsic mitotic growth rates, thus lending support to this view (Da Silva and Bell 1992).

On the other hand, sexual and natural selection might reinforce each other and so promote phenotypic and ecological divergence (Rice and Hostert 1993; Rundle and Nosil 2005). Although sexual and natural selection might work together in many different ways (reviewed in Schluter 2000a), the key distinction is which process is the main driver of diversification. If it is sexual selection, then chance initial differences in mate preferences may become exaggerated by antagonistic coevolution between the sexes, which can lead to divergence in mate preferences that are fortuitously, but not causally, associated with ecological differences. On the other hand, if ecological divergence is occurring already, then sexual selection can promote phenotypic diversification by reducing gene flow among habitats if mating is more likely to occur among males and females from the same habitat. Such assortative mating could itself have different causes. Antagonistic coevolution driven by sexual conflict could again be important, especially if the costs of sexual display vary across habitats, or if different traits happen to be preferred in distinct habitats (called sensory drive; see, e.g., Boughman 2001). It's also possible that sexual selection might align with natural selection to aid adaptation to a novel environment if mating success is correlated with fitness under natural selection—the so-called good genes model of sexual selection (Candolin and Heuschele 2008; Rundle, Chenoweth, and Blows 2006). This mechanism can promote the fixation of beneficial alleles and lead to more rapid elimination of deleterious alleles, and so increase the rate of adaptation to a novel environment.

Evidence from experiments with insects, mostly *Drosophila,* is mixed (reviewed in Arbuthnott and Rundle 2013). In microbial systems, currently only one test is known of the hypothesis that sexual selection and divergent natural selection, when occurring together, might lead to more pronounced ecological divergence. Bell (2005) propagated a genetically diverse population of *C. reinhardtii* capable of mating in liquid media and, unusually, on solid media for about five years, or roughly 100 sexual cycles and 1000 mitotic generations. In terms of the main hypothesis that coincident natural and sexual selection in different environments would lead to ecological divergence and speciation, the experiment was a failure: there was no evidence for the evolution of sexual isolation between the lines selected in the two environments. On the other hand, a radical change in the gender system occurred during the experiment. *C. reinhardtii* is normally heterothallic, meaning it can produce sexual offspring only with an individual of the opposite mating type. Homothallism, or self-fertilization, rapidly evolved during the experiment, however, and seems to be linked to the transposition of a mating-type gene to an autosomal chromosome.

What's important here is not this odd result, as intriguing as it is, but that we are still some way from understanding how sexual and natural selection interact to cause phenotypic divergence. More experiments are needed, as well as new experimental systems. One promising direction might be to exploit the chemical signaling involved in mate recognition of different yeast mating types (Rogers and Greig 2009), but no doubt other systems are waiting to be described. Expanding the range of microbial model systems available for studying the connection between sexual reproduction and diversification should thus be a priority for the field in the coming years.

CONCLUSIONS

Australopithecus sediba, the wellspring ape, remains something of an enigma. Not quite an ape, yet at the same time clearly not human. It's almost too easy to label it the "missing link" between tree-dwelling ape and upright, savanna-walking human.

Yet, as with most evolutionary riddles, the real story of how *A. sediba* came to look the way it does is probably more complex and at the same time more mundane. Complex because *A. sediba* looks like it does thanks to the multiple sources of selection that were at work. The abiotic environment must have played a role, but so too would competition from other early human species coexisting on the African plains. Mundane, at least to most evolutionary ecologists, because it's not so surprising that both the physical environment and the biotic environment should play important roles in driving evolutionary diversification. Why should humans be any different?

Convincing tests of the role played by competition in driving early human evolution will be difficult, however. When all we have to work with are a handful of fossils, the signal of competition's role is difficult to disentangle from other potential causes of diversification. Although no one disputes that it's a long way from the African plains of 2 million years ago to a test tube filled with microbes in a lab somewhere in the Americas or northern Europe, at least with microbes we can begin to separate the relative contributions of different factors to diversification.

This has been our goal with this chapter. We began by looking at a framework for understanding how four basic processes—mutation order and drift, epistasis, divergent selection, and ecological interactions—can contribute to evolutionary diversification in phenotypic form. These processes operate in both asexual and sexual systems, and so are very general. If our focus were to include sexual systems, we would need a fifth factor, sexual selection.

What have microbial systems told us about the role of each of these processes in fitness and phenotypic divergence? The findings can be summarized as follows:

1. The effects of stochastic processes associated with mutation and drift cause phenotypic disparity, measured both in terms of fitness and in terms of non-fitness phenotypes, to increase through time and decrease with increasing population size.

2. Epistasis, and in particular, reciprocal sign epistasis, can contribute to divergence.

3. Divergent selection invariably leads to ecological differentiation in the form of niche specialization; this can lead to reproductive isolation evolving as a by-product.

4. Phenotypic divergence caused initially by divergent selection may be further exaggerated by ecological interactions such as competition, parasitism, and predation, but no evidence to date suggests that mutualism plays a role in phenotypic diversification.

5. There is limited and mixed evidence about the role of sexual selection in diversification.

Sexual selection aside, a clearer picture of the relative contribution of these processes is starting to emerge. Together, mutation-order effects and epistasis contribute only modestly to ecological divergence; to confirm this view, treat-

ment effects due to environment invariably emerge as a statistically significant factor in any analysis of variance. Analysis also suggests that divergent
selection is almost always a far more important cause of diversification, at
least in laboratory experiments. The qualifier *almost* is needed here, because
investigators are just now getting together the tools to examine the prevalence
of sign epistasis in experimental evolution. Sign epistasis reflects underlying
constraints on the metabolic and biochemical machinery of cells, and it may
yet prove important in driving diversification. The increasing availability of
whole genome sequencing data, when combined with careful genetic analysis,
will tell us more in the near future.

Ecological interactions most effectively drive phenotypic diversification
when they work in concert with underlying divergent selection. Certainly this
is true for resource competition, which can and does often lead to character
displacement, even in microbial systems. It's also true for enemy-victim systems where diversification is often associated with a trade-off between resource
competition and parasite or predation resistance. Moreover, ecological interactions can drive diversification on their own when they generate antagonistic
coevolution, as we saw in the case of *P. fluorescens* and its phage. Divergence
into locally adapted populations in these situations will then be aided by
mutation-order effects and epistasis, although the importance of these factors
in contributing to the evolution of local adaptation in host-parasite systems
deserves more study.

Some notable gaps also need filling. Paramount among these is the role
of antagonistic coevolution in generating diversification and local adaptation. Antagonistic coevolution may be crucial for generating and maintaining diversity in natural systems, over and above the diversity generated by
divergent selection. It also has been suggested that antagonistic coevolution
is vital to explaining the evolution of sex, an idea known as the Red Queen
hypothesis (Bell 1982). The challenge at the moment is that very few examples
of antagonistic coevolution exist, and they are all restricted to host-parasite
systems.

There may be a good opportunity here to make progress by exploring the
various causes and consequences of antagonistic coevolution in different situations. Predator-prey coevolution is often thought to be important in driving
macroevolutionary patterns of diversification, yet we know little about this
process from an experimental perspective. Antagonistic coevolution between
the sexes, driven by the existence of different phenotypic optima for males and
females, is also emerging as an important component of sexual selection. If so,
then this may be an important source of ecological diversification in sexual
traits.

Exploring these ideas will require the development of new model systems.
In predator-prey systems, the challenge will be to find a model predator that

evolves quickly enough to keep up with adaptive changes in its prey. The favorite predators for many experimental microcosm studies are protists like *Tetrahymena,* but their use is complicated because they evolve slowly and have rather complex genetics. Studying sexual selection with microbes, on the other hand, requires the use of eukaryotic species such as *Chlamydomonas* or yeast. No doubt other, as yet unexplored, systems also exist. Developing these systems for use in the laboratory or in controlled field experiments will represent an important direction for future work on the evolution of phenotypic disparity and diversification.

CHAPTER 7

The Rate and Extent
of Diversification

There are few reasons to recommend flying through O'Hare International Airport in Chicago. It is one of the world's busiest airports. Storms, snow in winter and rain in summer, shut it down with almost tidal regularity. However, its one saving grace to the weary and frustrated traveler is the monumental skeleton of a 150-million-year-old brachiosaur keeping watch over the tarmac from inside Terminal 1.

It's hard not to be awed by this giant. The skeleton is 75 feet long and 40 feet high. Its head rises so high into the arched glass ceiling that it looks like it might break through. The neck is long and graceful, and the tail sweeps up

Figure 7.1 Mounted cast of *Brachiosaurus altithorax* in O'Hare terminal 1. Photo by Tristan Savatier.

and off the ground, seemingly weightless. The whole beast has an almost regal presence that lends an air of grace and dignity to the insult and stress of air travel.

Grace and dignity are not words often associated with dinosaurs. Extinction and catastrophe more commonly come to mind, at least among biologists and paleontologists. But these more refined descriptions remind us that the story of dinosaur evolution, diversification and demise involves some sense of majesty and no small amount of mystery.

Dinosaurs emerged from a single ancestral lineage about 230 million years ago following the end-Permian extinction, the largest extinction event Earth has ever known. Within 40 million years or so after dinosaurs first appeared, an unrelated competitor lineage, the "crocodile line" of archosaurs, or crurotarsans, had nearly gone extinct; the one lineage that survived eventually gave rise to modern-day crocodiles. Crurotarsan extinction opened up a wide range of new ecological opportunities for dinosaurs. They underwent a remarkable period of diversification, resulting in a marvelous array of species and forms. Giant sauropods like the one on display in Chicago, duck-billed hadrosaurs, the horned and beaked ceratopsids like *Triceratops,* and of course the ferocious-looking theropods like *Tyrannosaurus*—and my personal favorite for the Canadian content it brings to this book, *Albertasaurus.*

And then, 65 million years ago, dinosaurs all but disappeared, save for the lineage that eventually gave rise to birds. In with a flourish and out with a bang. You don't get much more majestic than that.

Beyond these few facts, we know little about how and why dinosaurs diversified as much as they did. Did dinosaurs cause the crurotarsan extinction, or were crurotarsans already on their way out and dinosaurs merely delivered the coup de grace? How important to the dinosaur's success were so-called key innovation traits like endothermy and an upright gait? What was the role of climate change, which was going from warm and wet to cooler and drier conditions, in the extinction of crurotarsans and the success of dinosaurs? Was dinosaur diversification fast or slow, in evolutionary terms? Did diversification happen all at once, or was it going on steadily throughout dinosaurs' history?

We likely won't get the answers to these questions any time soon. The fossil record is too spotty, especially early in dinosaur history. Moreover, gleaning useful data about the mechanisms responsible for diversification, things like competition or vacant niche space, is all but impossible when the remnants of skeletons are all we have to work with. Ideally, we would be able to go back in time and watch what happened ourselves (using imaginary rewind and fast-forward buttons to allow us to get the work done in the space of a typical grant cycle). Then we could construct experiments that would allow us to distinguish, for example, between competition from crurotarsans versus climate change as the major driver of dinosaur diversification.

Microbes, with their short generation times and their ability to be stored and revived at will, can be the rewind and fast-forward buttons we are looking for. First we must be willing to accept microbial systems for what they are—a representation of the more complex organisms and the more natural setting where diversification occurred. Then we can begin to explore the factors governing the rate and extent of diversification. Microbial work can thus provide needed experimental checks on the various theories and explanations that have been put forward to explain macroevolutionary patterns of diversification.

This chapter reviews the microbial experimental evolution literature for what it tells us about the factors governing the rate and extent of diversification. In this chapter, diversification refers to the number of ecologically distinct descendent types or species, not the degree of phenotypic divergence between them (the subject of the previous chapter). We start with a brief discussion of the process of diversification and the different approaches that have been taken to studying it. Then we consider the leading theory for diversification, what has come to be known as the ecological theory, and its key predictions. Most of the chapter is devoted to a discussion of direct tests of these predictions that come from microbial systems and the insight they give us into some of the more enduring debates in diversification studies. What emerges is an empirically robust theory of diversification that's intended to help clarify thinking about such long-standing debates such as the sources of ecological opportunity, the existence of ecological limits to diversification, and the various factors controlling the rate of diversification. The progress made in evaluating this theory with microbial experiments thus gives us a more complete view of how the diversification of all life, including dinosaurs, might have occurred.

THEORIES OF DIVERSIFICATION

Diversification is a hierarchical process. It starts when a population becomes differentiated into genetically distinct subpopulations and ends when those subpopulations become reproductively isolated from each other. Genetic differentiation at lower levels of biological organization thus forms the basis upon which all further differentiation accumulates at higher taxonomic levels. The result, over long time periods, is the branching pattern we see in phylogenies.

Theories of diversification seek to explain the variation among lineages and geographical locales in how branching occurs in phylogenies. More specifically, these theories seek to provide explanations that can account for variation in the rate—how rapidly branching occurs—and the extent, meaning the number of distinct genotypes or species (the tips of a phylogenetic tree), of diversification. The level of biological organization being explained differs according to the interest of the researchers and the availability of data. Paleontologists and comparative biologists have taken what might be considered a top-down approach by focusing on taxonomic scales at the species level and above. Evolutionary geneticists have come at the problem from the other direction, focusing on genes within populations and asking how effective selection or other forces are at generating genetic differentiation that leads to speciation. Speciation is thus the sole point of contact between the two approaches and so provides a natural starting point for this discussion.

Theories of speciation have traditionally been classified on the basis of geography, to distinguish between speciation happening through geographic isolation (allopatry) or not (sympatry). More recently, an effort is being made to refocus attention on the mechanisms causing reproductive isolation itself (Rundle and Nosil 2005). Theories of speciation differ in what they see as the ultimate causes of genetic differentiation and speciation. The range of possible causes is many and includes both natural and sexual selection, genetic phenomena such as increases in chromosome sets (polyploidization) or mutation-order effects (*mutation-order effects* refers to divergence associated with the order in which substitution of different alleles or genes occurs when evolutionarily independent populations adapt to the same environment; Coyne and Orr 2004; Schluter 2000a, 2009), and chance events associated with genetic drift or population bottlenecks.

Perhaps the two most general explanations for speciation proposed to date are theories based on divergent selection and mutation-order effects (Schluter 2009). The essential difference between the two is straightforward: those based on divergent selection treat divergence and speciation as the direct result of adaptation to different environments or niches, whereas those involving mutation-order effects see reproductive isolation arising from epistatic interactions among beneficial mutations substituted during adaptation to a single environment. So, for theories based on divergent selection, diversification is intrinsically tied to environmental variation but is largely (though not entirely) independent of it in mutation-order theories.

This chapter, like the book in general, emphasizes theories of diversification based on divergent selection. The reason for this is twofold. First, macroevolutionary studies of diversity often find a close relationship between diversity and various measures of environmental variation or environmental quality (Benton 2010; Ricklefs 2006; Rosenzweig 1995). This is especially true when it comes to explaining adaptive radiation, the rapid and often spectacular diversification of a lineage into a range of ecologically distinct daughter species (Schluter 2000a), where mutation-order effects fail as an explanation for the often striking phenotype–environment correlations that are their hallmark characteristic. Second, as we saw in Chapter 6, although mutation-order effects—meaning those derived from epistasis among beneficial mutations—can indeed cause phenotypic differentiation, the extent of divergence is usually much less than for that associated with divergent selection.

The Ecological Theory of Diversification

Ecological diversification occurs when divergent selection drives phenotypic differentiation. This process is common to both asexual and sexual species. In sexual species, the result can be the evolution of reproductive isolation and speciation, a process referred to as ecological speciation. Because this book deals primarily with asexual systems in the laboratory where recombination is not allowed, the term *diversification* is used here in the broader, more inclusive sense of ecological diversification. The term *speciation* refers to sexually reproducing taxa only.

In its most basic, stripped-down form, the ecological theory of diversification involves three components in roughly the following order: (1) divergent selection leading to niche specialization; (2) interactions among individuals driven either by ecology or sexual selection that, if present, generate disruptive selection and promote phenotypic divergence; and (3) the evolution of reproductive isolation or other forms of genetic isolation that prevent gene flow among lineages. This theory is attractive because it is very general. It applies with equal force to animals, plants, or microbes; requires no specific assumptions about ploidy or other features of the life cycle, nor about the geography of speciation; and is entirely compatible with both allopatric and sympatric speciation.

The ecological theory can make some explicit predictions, including how much diversity should evolve and be supported in an environment, the characteristic time course for diversification, and how quickly diversification happens. In the simplest scenario, we start with a single, ancestral lineage that has gained access to a novel set of conditions where competitors or enemies are absent. How extensively this lineage diversifies, measured as the number of descendant species that can be maintained at equilibrium, is determined ultimately by the number of niches in the environment. This prediction is a natural extension of the competitive exclusion principle, which says that no more types can coexist than there are limiting factors in the environment (Levins 1968). Thus the more niches there are, the more types can be supported (**Figure 7.2a**).

How diversity accumulates over time ultimately depends on whether niche space is fixed or variable (**Figure 7.2b**). If there is a defined, limited number of niches, then diversity dynamics will be characterized by an S-shaped curve: it increases rapidly early on when there are many empty niches but slows as niche space becomes filled (Benton and Emerson 2007; Sepkoski 1978; Rabosky and Lovette 2008; Raup et al. 1973; Rosenzweig 1995; Walker and Valentine 1984). The dynamics of diversity in this situation are said to be density dependent, where density here is understood to refer to the number of types, not population size. But what if niche space is never effectively filled,

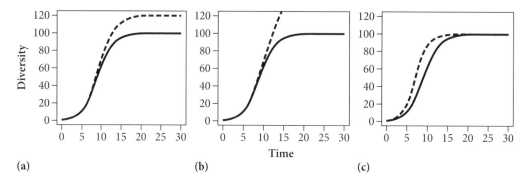

Figure 7.2 Predictions of the ecological theory of diversification. These predictions assume a logistic model of diversification with two parameters: the maximum rate of diversification, *r*, and the number of available niches, *K* (see text for further explanation). (a) Equilibrium level of diversity is higher when *K* is higher (dashed line compared to solid line). (b) Dynamics of diversification are S-shaped if ecological opportunity remains constant over time (solid line) but become J-shaped if new species create more niches (achieved by making *K* a linear function of *S,* the number of species; dashed line). (c) Disruptive selection caused by ecological interactions like resource competition is expected to increase the rate but not the extent of diversification (dashed line compared to solid line).

either because whatever limit exists is never reached or because new niches are continually created? Then diversity will increase exponentially and indefinitely, giving a J-shaped diversity-time curve. If reproductive isolation evolves alongside ecological specialization, then the rate of diversification is just the rate at which niche specialists evolve (see Chapter 2). This may not always be the case, because reproductive isolation is not always tied directly to ecological diversification. In that situation one ideally would track rates of speciation separately from rates of ecological specialization, although in practice it's not always possible.

Finally, different kinds of constraints can alter the rate of diversification by changing rates of speciation (or, equivalently, origination if the concern is with taxonomic levels other than species; **Figure 7.2c**). Genetic constraints, which make it harder for genetic and ecological isolation to evolve, slow rates of diversification. Alternatively, ecological interactions such as those associated with competitors or predators can either increase or decrease rates of diversification, depending on how they exert their effects. Intraspecific resource competition, for example, is often thought to generate disruptive selection and lead to more rapid rates of diversification. On the other hand, interspecific competitors often occupy what otherwise would be available niche space and so can prevent or slow diversification.

The ecological theory thus provides a potentially simple and compelling way to understand evolutionary diversification. Any factor that increases the extent of ecological opportunity, whether it is through a competitor's demise or the creation of new habitable area due to climate change, creates conditions that favor diversification. Any factor that reinforces and promotes divergent selection, like resource competition, will increase the rate of diversification; those that frustrate it, like gene flow, will slow it down. These two processes together, ecological opportunity and divergent selection, thus constitute the two basic ingredients required to generate diversity.

TESTING THE ECOLOGICAL THEORY OF DIVERSIFICATION

Direct, experimental tests of the ecological theory of diversification are straightforward to do in microbial systems as long as the researcher can identify and quantify ecologically distinct genotypes. Chapter 1 introduced the model adaptive radiation in *Pseudomonas fluorescens* as an example, where strikingly distinct colony morphologies associated with the evolution of niche specialization evolve over a few days in a static microcosm of nutrient-rich medium. The process of diversification is relatively easy to follow in this system because it's possible to count the abundance of different colony morphs through time. Other microbial model systems can include other kinds of phenotypes that appear to be associated with different ecological functions. Colony size variants are the most conspicuous and, as a result, are one of the most common polymorphisms studied.

Ultimately, though, these phenotypes are proxies for niche specialization. The diversity of niche specialist genotypes is what we are really interested in, and most experiments almost surely yield more genetic variation in a population that goes unnoticed if we rely solely on phenotypes. It's possible to describe variation in the response to different environments more directly by using a single, population-level variable called genotype-by-environment (GxE) interaction variance for fitness. To estimate GxE, one isolates individual colonies at random from a population and tests their fitness across a range of conditions. If all colonies isolated are of the same genotype, then there will be no GxE variance for fitness. On the other hand, if the population contains a diverse collection of genotypes that are specialized to different conditions, then GxE variance for fitness will be high. GxE variance can be further broken down into what has come to be known as responsiveness—variation in the amount of genetic variation expressed in an environment—and inconsistency, the change in rank order of fitness among genotypes across environments (Kassen and Bell 2000; Robertson 1959; Venail et al. 2008). When inconsistency exists, then different genotypes are fittest in different environments, and this is a necessary (though not sufficient) condition for divergent selection to maintain diversity. The GxE variance for fitness, and in particular its inconsistency component, is thus the most direct estimate of the amount of niche-specific genetic diversity maintained in a population and is therefore the preferred estimate of ecological diversity in microbial systems. Where this is not possible, proxy measures can often be useful, but always remember that such measures will mask an unknown amount of cryptic genetic and ecological diversity that is not revealed by the particular phenotype being studied.

HOW MUCH DIVERSITY?

The ecological theory predicts that the extent of lineage diversification should be a function of the range of ecological opportunities, that is, the number of distinct niches available to it. The logic behind this prediction stems from basic ecological and population genetics theory for the coexistence of genotypes or species. As a general rule, no more types can coexist than there are niches in the environment (Levene 1953; Levins 1968). This rule is not hard and fast: it can be shown that when resources fluctuate through time, more types can coexist than there are limiting factors in an environment (Armstrong and McGehee 1980; Chesson 2000). And organisms themselves can often create niches for others—just think about how a beaver dam changes the ecology of a forest stream. Still, when predicting the extent of diversification, the one niche, one type principle is generally a useful place to start.

Testing the prediction is another matter, because it's usually impossible to know in advance just how many niches are available to be occupied. Moreover, many organisms find ways of diversifying that at first glance seem unexpected but in retrospect make entirely good sense. Bacteria in continuous culture tend to stick to the walls of the culture vessel, for example. This tendency is frustrating to an experimenter interested in studying a physiologically and genetically uniform population, but from the bacterium's perspective, it's an eminently reasonable strategy to avoid being washed away with the waste. The lesson here is that ecological opportunity is both a property of the environment itself and something that can emerge from the activities of organisms. In other words, ecological opportunity is not a fixed property of the environment, but something that can itself evolve.

The tension between these two interpretations of ecological opportunity finds its way into many discussions of diversification. For example, in paleontological studies it's widely accepted that increases in habitable area, such as through geological processes or climate change, are often associated with increases in diversity (Benton 2009). Does this pattern come about because more physically distinct niches are available, or because there are more different ways to subdivide existing niche space through interactions with species that are present already? The answer from paleontology has not been clear. Microbial studies have a chance here to shed some light on how different sources of ecological opportunity contribute to diversification. In the following subsections, we consider three lines of evidence—post hoc explorations of the causes of diversification, direct manipulations of ecological opportunity, and ecological release experiments—that collectively clarify both the sources of ecological opportunity and its role in causing diversification.

The Sources of Ecological Opportunity

Is ecological opportunity created from abiotic properties of the environment, for example, the number of distinct resources or limiting nutrients? Or, does it arise from biotic sources associated with the activities of organisms themselves? According to the results of microbial experiments, the answer to both questions is, perhaps predictably, yes.

The abiotic environment often acts as a source of ecological opportunity in microbial experiments, as evidenced by the many examples of diversification in multiple resource mixtures of defined media (Barrett, MacLean, and Bell 2005; Blount, Borland, and Lenski 2008; Friesen et al. 2004; Hall and Colegrave 2007; Jasmin and Kassen 2007a). *Defined* here means "a medium where the different resources are added in known amounts." Diversification is also commonly observed in commercially available undefined media, which are often nutrient-rich digests of plant or animal material (Finkel and Kolter 1999; Habets et al. 2006; MacLean, Bell, and Rainey 2004). Imposing spatial structure onto resource heterogeneity can facilitate diversification by creating resource gradients and limiting migration among subpopulations (Habets et al. 2006; Korona 1996; Rainey and Travisano 1998). Spatial structure might be especially important for generating diversity in microbial populations forming biofilms, the aggregations of largely nonmotile bacteria like the plaque on our teeth or the scum that forms on the surface of spoiled milk (Poltak and Cooper 2011; Ponciano et al. 2009), and in spatially heterogeneous, resource-rich environments like the mammalian gut (De Paepe et al. 2011; Giraud et al. 2008).

There is also good evidence that biotic factors can act as sources of ecological opportunity in microbial systems. Most often, new ecological opportunities emerge because one genotype creates a niche for another. The classic example of niche creation in microbial systems is a cross-feeding polymorphism: one genotype produces a metabolic waste product that another genotype can use as a resource (Elena and Lenski 1997; Helling, Vargas, and Adams 1987; Rosenzweig et al. 1994; Rozen and Lenski 2000; Treves, Manning, and Adams 1998; Turner, Souza, and Lenski 1996). The evolution of resistance to shared enemies is another example. If a trade-off between competitive ability and enemy-escape strategies exists, this can lead to the evolution and coexistence of multiple prey or hosts (Abrams 2000; Doebeli and Dieckmann 2000).

Such a result has been observed in phage–bacteria interactions (Brockhurst, Buckling, and Rainey 2005; Brockhurst, Rainey, and Buckling 2004; Schrag and Mittler 1996) and predator–prey interactions (Gallet et al. 2007; Hall, Meyer, and Kassen 2008; Meyer and Kassen 2007).

Moreover, biotic activity can create new ecological opportunities in another way: indirectly as a result of adaptation and resource competition. Spencer et al. (2008), for example, showed that populations of *Escherichia coli* propagated in a mixture of glucose and acetate eventually diversify into coexisting glucose-specialist and glucose-acetate generalist subpopulations, but this effect occurred only after a period of about 400 generations involving increasing specialization on glucose. Presumably, increasing specialization on glucose drove down glucose concentrations to such a low level that it became worth incorporating acetate, which is a non-preferred resource for *E. coli* when glucose is present, in a manner resembling phenotypic divergence through character displacement (Tyerman et al. 2008). Similar dynamics, involving an initial period of adaptation followed by diversification, have been observed in another *E. coli* experiment, although here the mechanisms supporting diversity are less clear (Maharjan et al. 2006).

Direct Manipulations of Ecological Opportunity

The extent to which a lineage diversifies (i.e., the number of distinct daughter lineages) depends on the range of ecological opportunities available: more ecological opportunity should mean more diversity (Benton and Emerson 2007; Sepkoski 1978; Simpson 1953; Walker and Valentine 1984). There are a handful of direct tests of this prediction. The most compelling is the *P. fluorescens* radiation into a range of niche specialist types that are associated with oxygen availability, as described in Chapter 1 (Rainey and Travisano 1998). Ecological opportunity in these experiments comes about because a static microcosm, one that remains undisturbed in an incubator, generates spatial structure in the form of an oxygen gradient. Oxygen diffuses into the broth from the air and rapidly becomes limiting as the population size of the ancestral genotype increases. As the ancestor's population expands, mutation introduces genotypes capable of forming a biofilm at the air–broth interface, and this affords them the ability to take advantage of the abundant oxygen there. So the emergence and maintenance of diversity in *P. fluorescens* requires spatially structured microcosms. Ecological opportunity can be removed by agitating the microcosm continuously on an orbital shaker, which redistributes oxygen throughout the medium and destroys any oxygen gradient that would otherwise form. Without the oxygen gradient there is no longer any spatial structure, and diversity is lost from previously diverse populations or fails to evolve from genetically uniform ones (Buckling et al. 2000; Kassen et al. 2000; Rainey and Travisano 1998; Rainey and Rainey 2003).

Ecological opportunity has been manipulated in other ways. Colonies growing in spatially distinct areas on the surface of an agar plate create gradients in resources such as metabolic waste products that leak out of the cell. These waste products act as resources for other genotypes through cross-feeding. Habets et al. (2006), for example, showed that long-term selection of *E. coli* colonies whose spatial position on a plate was maintained for close to 1000 generations led to the emergence and maintenance of diversity. However, destroying the spatial arrangement of colonies by washing them off the plate and then reinoculating from the mixed population of cells did not have the same effect.

The availability of distinct resources is also a ready way of adjusting ecological opportunity. Barrett, MacLean, and Bell (2005), for example, observed higher diversity after nearly 1000 generations of selection in well-mixed environments containing multiple carbon substrates as compared to single-substrate environments, although the relationship between the number of substrates and the extent of diversity was by no means direct. In fact, diversity was more likely to evolve when multiple substrates were available, but more diversity did not necessarily evolve when there were more substrates. This is a common observation: selection in resource mixtures often, though not always, leads to higher diversity than in single-substrate environments (Bailey and Kassen 2012; Hall and Colegrave 2007; Jasmin and Kassen, 2007a, 2007b). In these experiments, the evolution of diversity is often sensitive to the substrates chosen and how the population uses them. If there is preferential use of one substrate over another for physiological reasons (Görke and Stülke 2008), or because one substrate is exceptionally high yielding, adaptation can be biased toward the preferred or high-yielding substrate and diversity may fail to evolve (Bailey and Kassen 2012; Jasmin and Kassen 2007a).

Ecological Release Experiments

Diversification is often observed to take place when the normal suite of competitors that co-occur with a lineage are absent; this phenomenon is known as ecological release (Yoder et al. 2010). The radiation of dinosaurs following the extinction of crurotarsans is often cited as a classic example. The mechanism underlying this phenomenon is very simple. Competitors occupy niche space that otherwise would have been available to a focal lineage, so in their absence, the focal lineage is left with abundant ecological opportunity that spurs diversification.

Some direct tests of this mechanism have been conducted with the *P. fluorescens* radiation. Both broth-colonizing *smooths* and biofilm-forming *wrinkly-spreaders* will diversify if the alternative niche specialist is absent, suggesting that the absence of these competitors opens up new ecological opportunities. Consistent with this interpretation, the presence of a distinct niche specialist competitor in these experiments can prevent diversification. When biofilm-forming *wrinkly-spreaders* are introduced before a broth-colonizing smooth, for example, they prevent diversification (Brockhurst et al. 2007a; Fukami et al. 2007). Moreover, if two lineages that are both capable of diversifying are also competing in a novel environment, the one that diversifies first tends to exclude the other because it occupies available niche space (Bailey et al. 2013).

Ecological release can also occur when a lineage escapes a predator. The idea here is that the predator keeps prey numbers low enough to prevent resource competition from driving diversification. Once the predator is gone, however, prey numbers increase again, leading to intense resource competition and diversification. No explicit tests of this idea are known, but as discussed in more detail later, there is good evidence that predators can depress prey numbers enough to slow or even prevent diversification in the *P. fluorescens* radiation (Buckling and Rainey 2002b; Meyer and Kassen 2007).

DYNAMICS OF DIVERSIFICATION

Diversification occurs because a single ancestral type gives rise to two daughter types. Daughter types can also go extinct, so the net rate of diversification is typically modeled as the difference between rates of origination (λ; or rates of speciation, if we are talking about sexual species) and extinction (μ). There is a natural parallel here with models in population biology that track birth and death rates of individuals. So perhaps it's no surprise that models for the dynamics of diversification bear a close resemblance to those for population growth (Sepkoski 1978).

The result is that two distinct dynamic patterns of diversification have been proposed (Benton 2009; **Figure 7.2b**). The first is the expansionist model, where diversity (S) increases exponentially over time: $dS/dt = rS$. The intrinsic rate of increase, r, is the difference between the rate of origination (λ) and the rate of extinction (μ). Diversity increases without limit either because no limit exists, perhaps because each new species creates niches for other species (diversity begets diversity), or because a limit exists but in practice is never reached. This may be because real limits on diversity are simply very high, or because periodic mass extinction events reduce total diversity to such a low level that the recovery phase is effectively exponential. The dynamics of this process lead to a J-shaped curve of diversity over time.

The second is the saturating or equilibrial model, where an upper limit to diversity is set by available niche space. This naturally leads to the dynamics of diversity being modeled as a density-dependent process, as in logistic models of population growth: $dS/dt = r_0S(1 - S/K)$. Now, r_0 represents the difference in origination and extinction rates when diversity is small, and K is a "carrying capacity" for diversity that in principle is set by the number of niches available in the environment. As diversification proceeds, ecological opportunities become filled and the rate of diversification slows due to declining origination rates and higher extinction rates. This model leads to an S-shaped relationship between diversity and time.

Which model is a better descriptor of diversity dynamics? The fossil record doesn't provide a clear picture, because both models show similar dynamics when diversity is initially low relative to available niche space. Periodic reductions in diversity due to mass extinctions or a secular trend toward niche expansion driven by, for example, global changes in climate or habitable area, can thus both lead to expansionist diversity dynamics (Benton 2009; Benton and Emerson 2007; Stanley 2007).

Phylogenetic approaches are not much help either. Because phylogenies record only speciation events that have survived to the present and not extinctions, they always give the impression that diversity increases through time. Distinguishing among alternative models of diversification is fraught with challenges because a given branching pattern is often consistent with different models of diversification. Even more worrisome is that the ecological context in which diversification occurred is almost always unknown. Did diversification begin with a single ancestor that gained access to a range of new niches in the absence of competitors or predators, as in the basic models described earlier? Or was diversification occurring in a community that was already saturated with species, so that new species simply replaced old ones? Ecological information is not easily recorded in a phylogeny, so it's impossible to tell. We do know that in many phylogenies there is a signal of lots of short branching events near the base and long terminal branches toward the present (McPeek 2008; Morlon et al. 2012; Rabosky and Lovette 2008). This pattern is consistent with diversification being faster early in the history of many clades compared to later, as the equilibrial model predicts. Nevertheless, distinguishing in a convincing way among alternative models remains a challenge. These and other problems associated with inferring patterns of diversification from phylogenies are discussed in more detail by Losos (2011) and Ricklefs (2007).

Evidence from Microbial Experiments

What we know about the dynamics of diversification in microbial experiments comes almost exclusively from work with *P. fluorescens*. The diversity-time relationship in static microcosms seeded with the broth-colonizing *smooth* founder strain is roughly S-shaped, consistent with the equilibrial model, although a noticeable peak of diversity occurs after approximately four days and is followed by a slow decline (**Figure 7.3**; Fukami et al. 2007; Meyer et al. 2011). This pattern, called overshooting, is often seen in many "real" adaptive radiations (Foote 2007; Gavrilets and Vose 2005; Schluter 2000a). The causes

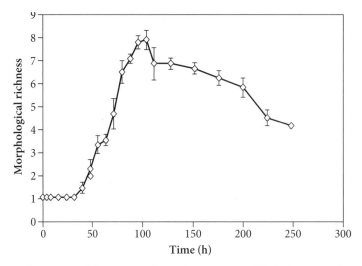

Figure 7.3 Dynamics of diversity in the *P. fluorescens* model of adaptive radiation. Data is the mean colony morphotype diversity (± 1 standard error) for three independently sampled microcosms over time in hours. Adapted from Meyer et al. (2011).

of the decline phase are not well understood in general. In the *P. fluorescens* radiation, we have shown that decreases in diversity are due to the selective replacement of genotypes within a niche class and not to wholesale losses of niches themselves (Meyer et al. 2011).

The S-shaped pattern appears to be quite robust in *P. fluorescens* experiments. We see it when the founding strain is co-cultured with an otherwise isogenic strain that cannot diversify (Bailey et al. 2013). Zhang, Ellis, and Godfray (2012) saw a similar pattern in the same system when the competitor was a different species (*P. putida*). The addition of predators (Meyer and Kassen 2007) and the timing and frequency of disturbances (Massin and Gonzalez 2006) also have little impact on the dynamics of diversity in this system. These results suggest that there are real ecological limits to diversification in this system.

A handful of other microbial studies have followed the dynamics of diversity through time. Most of them show an S-shaped dynamic, or at least decelerating diversification rates as the experiment proceeds (Korona 1996; Maharjan et al. 2006; Stevens et al. 2007). Only one experiment shows a different pattern. Barrett and Bell (2006) found that diversification rates remained relatively constant through time when *P. fluorescens* was allowed to adapt to a range of different resources over 1000 generations. This effect seems to have occurred because few genuine niche specialists had evolved at the end of the experiment and because most genotypes are fairly broadly adapted to the available resources. Thus diversity may still be increasing because niche specialization was still evolving. No known experiments have observed exponential diversity dynamics that would be consistent with a diversity-begets-diversity mechanism for diversification. Nevertheless the many examples of cross-feeding polymorphisms mentioned already are abundant evidence that diversity can sometimes lead to more diversity. However, niche creation seems to contribute only modestly to the dynamics of diversity in the experiments studied to date.

RATES OF DIVERSIFICATION

Diversification rates can vary through evolutionary time if origination and extinction rates are diversity dependent, as we just saw. Rates of diversification are also known to vary markedly across both taxa and environments (Sepkoski 1998). At one extreme, for example, are groups of living fossils that exhibit extremely low rates of diversification and have persisted nearly unchanged for long periods of evolutionary time (e.g., gingko trees: Royer, Hickey, and Wing 2003). At the other are adaptive radiations, where rates of diversification, and in particular speciation, are typically very high (Losos 2010). Little is known about what controls this variation, although many guesses and suggestions have been made.

Proximately, the cause of variation in diversification rates comes from either higher speciation/origination rates or lower extinction rates. A large body of literature aims to identify the causes of rate variation from phylogenies (Barraclough, Harvey, and Nee 1995; Harmon et al. 2003; Nee 2006; Rabosky and Lovette 2008). To quickly summarize this literature, rate variation is attributed to properties of either the species or group being considered (the presence or absence of a special trait or key innovation, for example) or to the environment that group occupies (heterogeneity, area, energy, ecological opportunity). In paleontology, these two views are often referred to as the Red Queen and Court Jester models of evolutionary change and diversification, respectively. Red Queen models are named for the Red Queen in Lewis Carroll's *Through the Looking Glass,* who tells Alice, "It takes all the running you can do, to keep in the same place." These models see biotic interactions that generate antagonistic coevolution among species as the main drivers of evolutionary change and diversification (Van Valen 1973). Court Jester models, on the other hand, take their name from the variable and whimsical behavior of the medieval court comedian (Barnosky 2001). They attribute rapid rates of diversification to changes in the physical environment that are little affected by the activities of organisms themselves—think of the meteorite that killed the dinosaurs as an example. Again, many methodological challenges are associated with inferring diversification rate variation, not least of which is trying to separate the relative contributions of changes in speciation as opposed to extinction rates to net diversification (Losos 2011; Ricklefs 2007). Let's now review what microbial diversification experiments have told us about the various factors that contribute to changing net diversification rates.

Serendipity

Paleontologists have long noted that diversification rates become exceptionally high only when conditions are, like the temperature of Goldilocks' porridge, "just right" (Benton 2009; Gould 1989; Sepkoski 1998). Defining what just right means, and predicting when those conditions will arise, is not always easy and makes diversification seem to happen rather serendipitously. Still, one clear pattern does emerge: diversification rates in the fossil record are often highest following changes in the physical environment associated with, for example, mass extinction events or the opening up of new habitable area through climate change and tectonic movements. This pattern suggests that the conditions spurring diversification are those that create new ecological opportunities. We saw earlier in the chapter that rates of diversification in microbial experiments are often highest when ecological opportunity is abundant, suggesting that the opening up of new ecological opportunities is a plausible explanation for high rates of diversification.

Genetic Constraints

A genetic constraint arises when a genotype is unable to generate appropriate genetic variation that allows it to adapt and diversify in a novel environment. Genetic constraint has two main causes. The first is when the genetic architecture of a population prevents or limits adaptation to a novel environment from standing genetic variance due to negative genetic correlations among traits (Blows and Hoffman 2005). A trade-off in fitness across environments implies a negative genetic correlation for fitness, meaning that almost no genetic variation is capable of increasing fitness in both environments simultaneously. An example comes from Bell (1997b), who showed that mean fitness did not increase when an initially genetically diverse population of the unicellular alga *Chlamydomonas reinhardtii* was selected in a spatially variable environment with dispersal, where each patch contained a different combination of limiting resources. Bell's result is consistent with Dickerson's (1955) suggestion that the genetic correlation required to prevent a response to selection in a patchy environment is $-1/(N-1)$, where N is the number of patches. The implication is that for diversification in a novel environment that requires adaptation to multiple niche dimensions, the genetic correlation in fitness across environments can prevent this from happening, at least in the short term. Over longer periods, mutation should be able to break down this constraint, but there are no known examples of this phenomenon.

A second cause of constraint derives from the inaccessibility of beneficial mutations due to changes in the mutation supply rate associated with adaptation. As a population nears an adaptive peak, the supply of beneficial mutations decreases (see Chapter 1) and makes diversification increasingly difficult. Consistent with this mechanism, *smooth* morphotypes of *P. fluorescens* that had adapted to the broth phase were unable to diversify into the usual suite of niche specialist morphotypes when cultured in static microcosms (Buckling, Wills, and Colegrave 2003). Epistasis can also limit access to beneficial mutations. This effect was shown by Woods et al. (2011), who demonstrated that genetic interactions among beneficial mutations reduced the accessibility of further beneficial mutations in high-fitness genotypes and allowed lower-fitness genotypes to sweep to fixation. However, no studies are known to have examined the effect of such genetic interactions on diversification itself.

Key Innovations

A key innovation is a trait or complex of traits that, once acquired, affords access to a novel environment or new ecological opportunities that in turn spur diversification (Heard and Hauser 1995). Commonly cited key innovations are the evolution of endothermy and upright gait in dinosaurs (Brusatte et al. 2010) and the evolution of phytophagy in insects (Mitter, Farrell, and Wiegmann 1988). Perhaps the most striking example of a key innovation in microbial experiments is the evolution of citrate use in one of the long-term *E. coli* populations that led to the stable coexistence of citrate-using and non-citrate-using genotypes (Blount et al. 2012). The inability to use citrate under aerobic conditions is a diagnostic trait of *E. coli*. After 31,500 generations of selection in a minimal medium containing glucose and citrate, however, a subpopulation within one line had evolved the ability to metabolize citrate, leading to the stable coexistence of citrate and glucose specialists. The evolution of citrate use could have been due simply to the chance occurrence and subsequent selection of an extremely rare mutant. But the observation that the chances of re-evolution of citrate metabolism decreases as one goes back in time is at odds with this explanation. In fact, citrate-metabolizing mutants were more likely to occur later in the experiment. This pattern strongly suggests that citrate metabolism required the prior substitution of a specific suite of mutations before evolving, and they must have evolved for reasons that only fortuitously allowed the evolution of citrate use. Other putative examples of key innovations in the microbial literature include a number of studies involving the evolution of novel substrate use (Bailey and Kassen 2012; Lin, Hacking, and Aguilar 1976) and novel routes to infection in phage (Meyer et al. 2012). In Chapter 9, we will discuss the details of how key innovations evolve.

Ecological Interactions

We saw in Chapter 6 how ecological interactions generated by competitors or antagonists like predators or parasites can either promote or impede phenotypic divergence through their effects on the strength of divergent and disruptive selection or the availability of ecological opportunity. For the same reasons, these same ecological interactions can affect the rate of diversification as well. Whether ecological interactions increase or decrease diversification rates depends on how they change the strength of divergent selection. Ecological interactions can increase rates of diversification if extreme phenotypes enjoy higher fitness than intermediate phenotypes, generating disruptive selection. On the other hand, ecological interactions may reduce rates of diversification by, for example, reducing population density and so weakening the strength of divergent or disruptive selection or by eliminating the ecological opportunities that cause divergent selection altogether.

What happens in microbial experiments? Consider competition first. Intraspecific competition is often held to be a major driver of diversification, because individuals share very similar resource requirements or niches that can generate strong disruptive selection (Doebeli and Dieckmann 2000; Schluter 2000a, 2000b; Simpson 1953; Yoder et al. 2010). A number of microbial experiments implicate resource competition as the mechanism driving diversification although, as noted previously, this is because alternative mechanisms such as predation or parasitism are excluded by design and not because the strength of resource competition has been manipulated directly (Brockhurst et al. 2006; Friesen et al. 2004; Habets et al. 2006; Jasmin and Kassen 2007a; Rainey and Travisano 1998; Tyerman et al. 2008). Fewer studies have examined the impact of competition on rates of diversification, although when they do the results largely match predictions of the theory. Reducing the density of a diversifying lineage should lead to weaker resource competition and so to slower rates of diversification, and results consistent with this have been observed when the density of a diversifying lineage of *P. fluorescens* is reduced by predators (Meyer and Kassen 2007) or parasites (Buckling and Rainey 2002b). The strength of resource competition experienced by a diversifying lineage might actually be increased by the presence of other genotypes that themselves cannot diversify. This happens because the non-diversifying competitor reduces resource abundance for the diversifying lineage and so generates what amounts to stronger disruptive selection. An example of this effect was recently reported by Zhang, Ellis, and Godfray (2012), who tracked the dynamics of diversification of *P. fluorescens* in the presence and absence of a competitor species, *P. putida*. Similar results have been reported in experiments with *P. fluorescens* when the standard diversifying lineage is co-cultured with isogenic genotypes that lack operons associated with biofilm formation (Bailey et al. 2013).

Interspecific competition, on the other hand, is often associated with reduced rates of diversification because the competitor species use substantially different resources or sets of resources, thereby eliminating ecological opportunity (Losos and Ricklefs 2009). The microbial literature includes just a few direct tests of this idea, all from work with *P. fluorescens*. All show unequivocally that niche occupation not only slows the rate of diversification, but often prevents it altogether (Brockhurst et al. 2007a; Fukami et al. 2007; Bailey et al. 2013).

Now consider shared enemies such as predators or parasites. Shared enemies can generate disruptive selection if intermediate phenotypes have low fitness because they are susceptible to predation or parasitism (Vamosi 2005). They can also create ecological opportunities in the form of predator resistance in prey (Vermeij 1994). In Chapter 6, we saw many examples of shared enemies driving diversification in microbial experiments by creating ecological opportunities (Bohannan and Lenski 2000; Brockhurst, Rainey, and Buckling 2004; Brockhurst, Buckling, and Rainey 2005; Buckling and Rainey 2002b; Chao, Levin, and Stewart 1977; Gallet et al. 2007; Lenski and Levin 1985; Meyer and Kassen 2007; Schrag and Mittler 1996). Only one such experiment has focused explicitly on the role of shared enemies in governing the rate of diversification. Meyer and Kassen (2007) found that a protozoan predator reduced prey density by about tenfold. The consequence was weaker diversifying selection, as evidenced by a shallower slope when fitness is expressed as a function of starting frequency relative to the absence of a competitor. The investigators interpreted this result to mean that the reduced prey density weakened resource competition, leading to weaker disruptive selection and delayed emergence of diversity in the *P. fluorescens* radiation. Thus two factors are at play in determining whether predators increase or decrease prey diversification rates: the balance between the creation of new ecological opportunity in the form of predator-free space, which increases diversification rates, and the degree to which prey population sizes are reduced by predation, which slows it.

It's also worth mentioning mutualism here, if only for completeness. Mutualism represents a form of coevolution that can in principle accelerate diversification rates in both species by creating new forms of interactions that either allow access to novel ecological opportunities (Margulis and Fester 1991). These new interactions can generate disruptive selection, as explained in Chapter 6, due to a trade-off between one species' ability to exploit its own resources and the benefit gained from the resources supplied by the mutualist (Doebeli and Dieckmann 2000). Currently, no experiments are known to have examined the impact of mutualism on rates of diversification.

Sexual Selection

Sexual selection can promote diversification by encouraging assortative mating within populations, which reduces gene flow among populations and can make divergent selection more effective at generating niche specialization. Although a number of comparative studies, beginning with the seminal paper by Barraclough, Harvey, and Nee (1995), have supported the idea that sexual selection increases rates of diversification, the evidence overall is less convincing (Ritchie 2007). Beyond the experiments mentioned in Chapter 5 that have focused on the role of sexual selection in promoting phenotypic divergence and adaptation, no explicit test that sexual selection per se can increase diversification rates is known. There is clearly an opportunity here for future work.

CONCLUSIONS

Dinosaurs, with their gigantic size and monstrous form, retain a certain majesty in our collective imagination. This majesty is reinforced by the sweep and flourish of their evolutionary history: rapid diversification early on, followed by an abrupt and perhaps even violent demise. Dinosaurs may have been one of life's grandest gestures, an experiment in form and lifestyle that played out over millions of years. But the processes that led to the dominance and diversity of reptilian life on land for over 200 million years were not themselves mysterious. Dinosaurs evolved and diversified in much the same way as any other group.

The leading explanation for how diversity evolves is the ecological theory of diversification. The theory says that diversification is expected to occur when ecological opportunity, in the form of available niche space in the environment, creates divergent selection that leads to ecological specialization. Any factor that increases the extent of ecological opportunity or reinforces and strengthens the efficacy of divergent selection at generating phenotypically and genetically divergent types can increase either the amount of diversity created or how rapidly it evolves, respectively. Thus the many factors that have been suggested to play a role in driving diversification—predators, competitors, climate change, and so on—can be understood as operating primarily through their effects on either the range of ecological opportunity or on how effectively divergent selection generates ecological specialization. No more sophisticated or mysterious explanations are necessary.

The results of microbial selection experiments are in line with this general framework. The importance of ecological opportunity in diversification comes out very clearly in many experiments. Perhaps the most compelling example is that ecological diversification in *P. fluorescens* happens only in static microcosms where the formation of a biofilm provides ready access to oxygen, a key limiting resource, but not in shaken microcosms where oxygen is more evenly distributed. Fewer experiments have studied rates of diversification, admittedly, but those that were done are compelling. Adding predators to static microcosms in the *P. fluorescens* system, for example, decreases the strength of divergent selection and so, consistent with the predictions of the theory, slows the evolution of diversity. So there seems to be no reason to question the ecological theory as an explanation of diversity.

Microbial experiments have provided other important results about the factors governing the dynamics of diversification. Their key insights are summarized here:

1. The extent of diversification is often directly related to the amount of ecological opportunity, which itself comes about through both the physical properties of the environment and the activities of organisms themselves.

2. The dynamics of diversification are S-shaped, not J-shaped, though there can be an internal "hump" arising from a rapid initial increase in diversity followed by a decline. The causes of the decline phase are not well understood.

3. The creation of ecological opportunities by the activities of other organisms either happens too slowly or adds too few niches to greatly influence the dynamics of diversification.

4. Genetic constraints can slow diversification rates by making it difficult for an evolving strain to access appropriate genetic variation.

5. Reducing the strength of resource competition, for example by introducing a predator that reduces the density of an otherwise diversifying prey population, can slow diversification rates.

6. Key innovations that provide access to a novel environment can evolve through mutation-order effects.

These results give researchers insight into how diversification proceeds when driven by divergent selection. At least over the time scales being studied here—admittedly on the order of tens to thousands of generations rather than the millions or more typical of paleontological studies—we can say that the ecological opportunity driving diversification is effectively static. The dynamics of diversification thus behave as if they were diversity dependent. Diversification is fastest early on, when ecological opportunity is abundant, and slows as niche space becomes filled. Mechanistically, this likely happens because mutation, which introduces new variants, is not limiting in most microbial experiments, but competition for existing niche space becomes stronger as ecological opportunities are filled. A parallel might be made here between mutation and origination on the one hand and competitive exclusion and extinction on the other. If so, then in microbial microcosms, origination rates remain relatively constant through time but extinction rates increase. Whether this is an appropriate interpretation of the diversity dependence of speciation and extinction rates in more natural systems remains to be seen.

A great advantage of the ecological theory of diversification as described here is its generality. The theory provides an explanation for why some diversification events like adaptive radiations appear to be so fast and so spectacular while others seem, well, more mundane. The unusually rapid and extensive ecological diversification that characterizes an adaptive radiation comes about because it is the product of a fortuitous combination of factors: abundant ecological opportunity combined with strong disruptive selection, often driven by ecological interactions like resource competition . Rapid diversification also requires that the appropriate genetic variation be accessible to the incipiently diversifying lineage. If any of these factors are not in place, the result is expected to be a much less spectacular diversification event, or even no diversification at all. We have only a few good examples of adaptive radiation—Galápagos finches, Caribbean lizards, and cichlid fish to name three—not because they're hard to identify, but because they are genuinely rare. Adaptive radiation occurs only when a whole bunch of elements are aligned just so: ecological opportunity, disruptive selection, and few genetic constraints.

We must, of course, use caution when making broad generalizations for how diversification occurs on the basis of what we know about adaptive radiation alone. It's possible that adaptive radiation is such a rare event that diversification occurs more commonly via less spectacular routes. The alternative view is that adaptive radiation, although rare, contributes so much of life's diversity that it is the major mode by which species originate. Microbial experiments cannot reveal the truth of either of these claims; that remains a matter for comparative work.

What about the enduring controversy in the paleontological literature over the existence, or not, of ecological limits to diversification? The answer from microbial experiments studying this question is fairly clear: at least within the confines of life in a test tube, diversity has ecological limits. Niche creation, moreover, does not seem to add sufficient niche space to fundamentally change the dynamics of diversification. So the idea that diversity begets diversity, although clearly a phenomenon that happens in both test tubes and in nature, does not seem to change the overall dynamics of diversification, at least in microbial systems. As a starting point for interpreting historical patterns of diversification in the fossil record or phylogenies, a logistic-like or saturating model of diversification thus seems more appropriate than an exponential-like model.

Throughout this book, you will find explicit comments about the caveats that go along with interpreting the results of microbial experiments. Here, when it comes to diversification, we need to be especially careful for two reasons. The first is that much of what we know comes from just a handful of model systems and is especially biased toward the *P. fluorescens* system because it's so easy to work with. Biologists desperately need more examples of diversification and adaptive radiation in microbial systems, so that we can evaluate the robustness of the theory. The second reason for caution is that most of our inferences come from a fortuitous link between colony morphology and niche specialization. Indeed, colony morph variation is what often tips off researchers that something more interesting is going on in their cultures. As a result, we are almost certainly only scratching the surface regarding the real amounts of diversity. No doubt much more genetic variation is lurking unnoticed in these cultures than we can measure. A shift to high-throughput techniques for quantifying genetic variation in fitness will be an important way forward in getting a more accurate picture of diversity in these sorts of experiments.

A final caveat is, of course, that these experiments are all done in asexual systems. Genetic isolation, in the strictest sense, evolves as soon as a mutation occurs. However, many of the examples of diversification that we often care most about, dinosaurs included, concern sexually reproducing organisms. It remains to be seen how far the conclusions we have come to and discussed here can go toward explaining diversification in these more complicated systems. In these efforts, the use of facultatively sexual eukaryotes (Dettman et al. 2007; Dettman, Anderson, and Kohn 2008; Bell 2005; Rogers and Greig 2009; Schoustra et al. 2010) to study adaptive diversification will be especially important and valuable.

CHAPTER 8

Adaptive Radiation

If there is one chapter in this book that doesn't need an introductory vignette, this is it. Adaptive radiations are some of the most striking and spectacular examples of evolutionary diversification in the history of life. Everyone has a favorite example. Darwin's finches in the Galápagos, cichlid fishes in African Rift and Nicaraguan crater lakes, and the emergence of mammals at the end of the Cretaceous are among the most charismatic and best-known examples. There may be many others: Nearctic freshwater fish like whitefish and sticklebacks (see **Figure 8.1** for an example); silverswords,

Figure 8.1 A canonical example of adaptive radiation: threespine sticklebacks (*Gasterosteus aculeatus*) in freshwater lakes from British Columbia, Canada. Rapid diversification into a limnetic form specialized for open water (top two individuals; the male is on top and the female below) and a benthic form specialized on nearshore habitats (bottom two individuals; female is above the male) from a common marine ancestor within the last 15,000 years. These fish are from Enos Lake, B.C. The recent introduction of crayfish into the lake has caused the species pair to collapse into a hybrid swarm, leading to the loss of diversity. Photo by Dolph Schluter.

tetragnathid spiders, Hawaiian fruit flies, and honeycreepers (in fact just about everything endemic to the islands of Hawaii); dinosaurs following the end-Permian mass extinction; mammals following the Cretaceous-Tertiary extinction of dinosaurs; the list goes on.

For such a popular subject, it's surprising how little we know. In part this may be because adaptive radiations often show such obvious links between morphological and ecological diversity—think of how the beaks of Darwin's finches seem so exquisitely adapted to the kinds of seeds the different species feed on, or the vast array of mouth and body morphologies associated with different inshore habitats and resources in cichlid fish—that we scarcely feel the need to go much further. It's almost too easy to see the hand of natural selection in shaping diversity in most radiations.

Another reason for our lack of knowledge is that we actually don't have many good examples of radiations. The scope of this problem may be hard to appreciate from a survey of the literature, which is immense. You might get the distinct impression that many researchers use the same criteria to identify an adaptive radiation as U.S. Supreme Court Justice Potter Stewart did in identifying obscenity: "I know it when I see it." The rigorous application of formal criteria for identifying an adaptive radiation—common ancestry, adaptive phenotype–environment correlation, and rapid speciation—leaves us with preciously few compelling examples in nature (Schluter 2000a). Strong inferences about what causes a radiation are thus difficult to make on the basis of comparative studies alone. The problem is compounded because, even in the few good examples of adaptive radiation we do have, direct experimental tests of alternative hypotheses are not always something that can be done easily. The result is that the theory of adaptive radiation remains largely without extensive or robust empirical support.

This situation has changed dramatically in recent years with the increasing emphasis on the use of microbial populations that, because of their small size and rapid generation times, make possible the construction of replicated, manipulative experiments to study evolution in the laboratory. In this chapter, we'll review the contributions this work has made to our understanding of the ecological and genetic mechanisms underlying adaptive radiation.

Adaptive radiation is the rapid diversification of a single lineage into a range of ecologically and phenotypically distinct niche specialists. Radiations stand out from more mundane instances of diversification because they are spectacular, in the sense that rates of diversification are thought to be unusually high and the "fit" between phenotype and environment striking. Because there's such an obvious and close connection between morphology and ecology in adaptive radiations, they are often held to be canonical examples of natural selection causing diversification. The suggestion has often been made that adaptive radiations represent the major mode by which life on Earth has diversified (Rainey and Travisano 1998; Schluter 2000a; Simpson 1953).

The ecological theory of adaptive radiation (Schluter 2000a) is the leading explanation for how adaptive radiation proceeds. The theory can be understood as a three-step process. The first step is divergent natural selection caused by environmental variation, which leads to the evolution of niche specialist types that, by definition, trade off fitness across alternative environments. The second step, which may coincide with the first, involves ecological interactions such as resource competition or predation that promote the evolution of niche specialization by increasing the rate and extent of phenotypic and ecological divergence. The third step is speciation, or the evolution of reproductive isolation. It should be apparent that these are the same steps that drive diversification more generally, as explained in Chapter 7. The difference is that an adaptive radiation involves an unusual alignment of factors—abundant ecological opportunity and strong ecological interactions (often resource competition)—that together generate disruptive selection causing exceptionally rapid diversification.

The ecological theory is a compelling explanation for adaptive radiation, but it is incomplete on at least two counts. First, and most obviously, it does not include genetics. How does phenotypic and ecological divergence happen at the genetic level? Are there stereotypical genetic changes that can explain adaptive divergence? How many genes are involved in adaptation—few or many? Answering these questions will tell us something about the more fundamental problem of how genotype maps onto phenotype and whether adaptive radiation involves something unusual or unique, as is often assumed to be the case, that distinguishes it from more secular instances of diversification.

Second, the theory does not offer an obvious explanation as to how radiations come to differ in the extent of phenotypic divergence and the number of species they contain. Radiations in phylogenetically distinct taxa often reach dramatically different endpoints, an observation that could potentially be explained by invoking different ecological opportunities available to distinct founding lineages. But replicate radiations, where the founding lineage and the range of ecological opportunities (vacant niche space) appear to be the same, also vary—and this is a much harder observation to reconcile with the existing theory (Gillespie 2004; Meyer 1993; Pinto et al. 2008).

Here we review the literature on experimental evolution in microbes to address three gaps in the theory of adaptive radiation: the paucity of direct tests of mechanism, the genetics of diversification, and the limits and constraints on the progress of radiations. The goal is to point the way toward the development of a more general theory of adaptive radiation, one that integrates the ecological theory with genetics to provide a mechanistic account of the causes and limits of adaptive radiation.

With microbes the central property of interest in adaptive radiation, the degree of niche specialization, can be measured directly through competition experiments or other measures of fitness (such as growth rates or carrying capacity in pure culture) across a range of environmental conditions. Moreover, these measures can be taken at different times, making it possible to ask questions about the dynamic progress of adaptive radiations. This chapter summarizes experiments that track the emergence and fate of diversity, measured as genotype-by-environment variance in fitness or a close surrogate (such as colony morphology), in environments containing multiple niches. The rationale is that these conditions closely mirror those thought to be most commonly experienced during the initial stages of adaptive radiation. Note that we will not consider experiments that document the emergence of molecular or physiological diversity that is not explicitly tied to genetic variation in fitness.

DIRECT TESTS OF THE ECOLOGICAL THEORY

The characteristic feature of adaptive radiation is the rapid evolution of ecological diversity. The mechanisms responsible are strong divergent selection, often supported by ecological interactions that generate disruptive selection and so promote diversification, in the presence of abundant ecological opportunity, or vacant niche space. Microbial experiments provide some of the most direct and compelling examples of how these processes contribute to driving rapid ecological diversification.

Divergent Natural Selection Caused by Environmental Variation

Ecological diversification occurs because selection is divergent, favoring different types in under different conditions. As we have seen already in Chapter 3, there are many examples of divergent natural selection leading to the evolution of niche specialization in microbial systems. These experiments typically involve adaptation to spatially distinct resources, and so it's not hard to see how niche specialization could evolve through the conventional process of selective substitution of genes with environment-specific effects.

Ecological Interactions Promoting Diversification

It is often thought that both the unusually high rate of diversification and the marked phenotypic divergence among niche specialists during an adaptive radiation derives from the combination of divergent and disruptive (i.e., selection against intermediate forms) selection generated by different kinds of ecological interactions, in particular competition and predation (Rueffler et al. 2006).

Resource competition is the most commonly cited mechanism for disruptive selection and has undoubtedly played a role in some hallmark cases of adaptive radiation (Losos 2010; Schluter 2000a). Many microbial selection experiments likewise attribute the emergence of resource specialists and diversity directly to resource competition (reviewed in Chapters 6 and 7). Remember, though, that resource competition is often cited as the driver of diversification in these studies because alternative mechanisms such as predation and parasitism are excluded by design. Direct tests of resource competition's role in promoting diversification are few.

A telltale sign that resource competition has caused diversification is the coexistence of niche specialists through negative frequency-dependent selection. In the *Pseudomonas fluorescens* radiation (Chapter 1), for example, a rare, broth-colonizing *smooth* morphotype can invade a population of biofilm-forming *wrinkly-spreaders*. The reverse is also true: rare *wrinklies* can invade a population of *smooths,* meaning that both types cannot be lost from the population. However, other interactions can also generate negative frequency-dependent selection. They include facilitation, where one type gains a fitness advantage due to the presence of another (Day and Young 2004), and density-dependent selection in different environments (Levene 1953). The demonstration that fitness is negatively frequency-dependent thus usually strongly implicates resource competition as a driving mechanism for diversification, but does not guarantee it.

Even so, more and more evidence from a variety of sources supports the idea that resource competition is often the causal mechanism of diversification in microbial systems. Investigators have found, for example, that there is a positive relationship between the number of evolved niche specialists in the *P. fluorescens* radiation and the density of the radiating lineage: lower stationary phase densities, and so presumably less resource competition, lead to less diversity (Bailey et al. 2013). Grazing by the protozoan predator, *Tetrahymena thermophila,* during the *P. fluorescens* radiation reduces prey densities, and so presumably the strength of resource competition, leading to reductions in the strength of negative frequency-dependent selection as well as slower rates of diversification (Meyer and Kassen 2007). Three other experiments, by Barrett and Bell (2006), Jasmin and Kassen (2007a), and Tyerman et al. (2008), have found examples of character displacement—more extreme phenotypic divergence among individuals in sympatry than in allopatry—when populations are propagated on mixtures of substitutable resources. In the first two of these experiments, individuals isolated from mixtures were more divergent in terms of their fitness on the component substrates than were those isolated from comparable single-substrate environments. The third experiment, by Tyerman et al., showed that the growth profiles of two ecologically distinct genotypes became more similar when allowed to evolve in isolation. These results are hard to explain without invoking resource competition as a driver of phenotypic diversification.

Shared enemies such as predators or parasites can increase the rate and extent of divergence if intermediate phenotypes have low fitness, because they are susceptible to predation or parasitism (Vamosi 2005). The literature includes many examples of shared enemies driving diversification in microbial experiments (Bohannan and Lenski 2000; Brockhurst, Rainey, and Buckling 2004; Brockhurst, Buckling, and Rainey 2005; Buckling and Rainey 2002b; Chao, Levin, and Stewart 1977; Gallet et al. 2007; Lenski and Levin 1985; Schrag and Mittler 1996). However, no examples are known to have explicitly demonstrated this mechanism. The available evidence points to shared enemies either reducing the strength of diversifying selection due to reductions in prey density that weakens resource competition (Buckling and Rainey 2002b; Meyer and Kassen 2007) or creating novel ecological opportunities in the form of enemy-free space (see next section).

Ecological Opportunity

Some researchers have proposed that the range of unused or underutilized niches available to a lineage, or ecological opportunity, sets an upper limit on both the number of species that evolve during a radiation and the extent of phenotypic divergence among them (Benton and Emerson 2007; Simpson 1953; Walker and Valentine 1984). Using the metaphor of an adaptive landscape—how fitness changes as a function of genotype or phenotype—ecological opportunity can be interpreted as the number of fitness peaks accessible to a lineage. The distance between peaks determines the expected amount of phenotypic divergence between descendant lineages. Fisher's two-optima model represents one way of formalizing this idea (Chapter 3), and the consequences of this view for the evolution of phenotypic disparity were discussed at length in Chapter 6.

We saw in Chapter 7 how ecological opportunity can lead to diversification in microbial microcosms. The sources of ecological opportunity may be abiotic, for example in the form of new habitats or underutilized resources, or biotic, such as when the presence of enemies like predators and parasites creates selection for access to enemy-free space or the evolution of enemy-resistance strategies. Three additional points are worth emphasizing for their insights into how ecological opportunity can be created in the context of adaptive radiation.

First, spatial structure in the environment, which can promote diversification by providing novel niches in the form of gradients in abiotic or biotic factors and reducing the extent of gene flow among patches, can be an important contributing factor promoting diversification. In the *P. fluorescens* radiation, diversification happens only when there is spatial structure in the form of static, undisturbed microcosms. Diversity is lost from previously diverse populations or fails to evolve from genetically uniform ones when spatial structure is destroyed by shaking the microcosms on an orbital shaker (Buckling et al. 2000; Kassen et al. 2000; Rainey and Travisano 1998). Habets et al. (2006) also showed that spatial structure, this time resulting from the spatial positioning of growing colonies on an agar plate, was important in supporting diversity through frequency-dependent selection, presumably due to the creation of resource gradients around the growing colonies. Diversity remained low when spatial structure was destroyed periodically by washing the colonies off the plate before reinoculation. Spatial structure is clearly not necessary for adaptive radiation, though. A number of experiments have documented diversification in the absence of spatial structure (Barrett, MacLean, and Bell 2005; Friesen et al. 2004; Hall and Colegrave 2007; Jasmin and Kassen 2007a; MacLean, Dickson, and Bell 2005), likely as a result of strong competition in complex media containing multiple resources.

The second point is that, consistent with what is often seen in more natural systems, the absence of competitors can create ecological opportunity by freeing up previously unavailable resources (Benton 1995; Simpson 1953). A lack of competitors may explain why morphological divergence is often faster on islands, where the usual suite of competitors is absent, than it is on the mainland (Losos and Ricklefs 2009; Schluter 2000a). Direct support for this idea comes from work by Brockhurst et al. (2007a), who showed that diversification in the *smooth* ancestor of *P. fluorescens* was limited by the number of distinct niche specialists (*smooth, wrinkly-spreader, fuzzy-spreader*) already established within a static microcosm. The more different kinds of niche specialists were present, the less the *smooth* was able to diversify.

Third, it's very clear that the activities of organisms themselves create novel niches that can drive diversification (Emerson and Kolm 2005; Erwin 2008; Odling-Smee, Laland, and Feldman 2003). This phenomenon, sometimes called niche creation, niche construction, or facilitation, is most familiar in microbial experiments through the emergence of cross-feeding polymorphisms where one type feeds on the metabolic waste products of another type (Helling, Vargas, and Adams 1987; Rosenzweig et al. 1994; Rozen and Lenski 2000; Treves, Manning, and Adams 1998; Turner, Souza, and Lenski 1996). Niche construction also seems important to the evolution of the *fuzzy-spreader* morph in the *P. fluorescens* radiation, because this morph emerges only after the biofilm-forming *wrinkly-spreader* morph has evolved. Moreover, the *fuzzy-spreader* cannot invade a population of *wrinkly-spreaders* unless the broth-colonizing *smooth* also is present (Rainey and Travisano 1998).

These examples notwithstanding, there is little evidence from microbial experiments that facilitation is capable of creating new niches at a high enough rate to be a major contributor to diversity during a radiation. In the *P. fluorescens* radiation, for example, there is no evidence that the rate of colony morph diversification continues to rise as the radiation proceeds, as would be expected if facilitation were a major source of new ecological opportunities (Meyer et al. 2011).

Ecological opportunity should also govern the extent of phenotypic divergence among evolved types. As we saw with the discussion of the Fisher two-optima model in Chapter 3, the more strongly contrasted two environments are, relative to the starting position of the wild type in a fitness landscape, the more different the optimal phenotypes in those environments are likely to be (Day 2000; Via and Lande 1985). Jasmin and Kassen (2007a) tested this idea by selecting initially isogenic populations of *P. fluorescens* in environments composed of all pairwise combinations of four substrates for approximately 600 generations. They showed that phenotypic divergence, measured as the quantity of genotype-by-environment interaction in fitness when the fitness of isolates is assayed on the component substrates, increased as the substrates became more different. This effect depended somewhat on the relative productivity of the component substrates: selection was most effective at generating adaptation to the more productive substrate. The implication is that although divergent natural selection may be strong, diversification may not happen, because demographic asymmetries among patches constrain adaptation toward the most productive niche.

GENETICS OF DIVERSIFICATION

The characteristic feature of adaptive radiation, the evolution of novel niche specialists, requires a population to overcome three significant genetic barriers. The first barrier is ensuring a viable population, because invasion of a novel environment is likely to involve dispersal of just a few, maladapted individuals. The second is the availability of genetic variation that selection can act upon to generate adaptive divergence. Small initial population size means that standing genetic variation is likely to be severely depleted, and consequently adaptation must rely on novel genetic variants introduced through mutation. The third barrier is the evolution of fitness trade-offs, which are required for divergent selection to generate niche specialization. The evolution of trade-offs depends on the cumulative pleiotropic effects of genes substituted in the novel environment. The rapid evolution of niche specialization characteristic of adaptive radiations must involve the substitution of genes with environment-specific fitness effects. If the pleiotropic effect of most substitutions to increase fitness across a broad range of conditions, on the other hand, rapid ecological diversification is harder to achieve.

Overcoming the Barriers: The Exaptation-Amplification-Divergence Model

Models for the evolution of new gene functions (Bergthorsson, Andersson, and Roth 2007; Francino 2005) are a starting place for thinking about the sorts of genes or genetic systems involved in overcoming the first two challenges—small population size and limited extant genetic variation—that a population faces when it invades a new environment during the initial stages of a radiation. Population viability is made possible by exaptation, where preexisting gene functions are co-opted to permit growth and reproduction under novel conditions. Increased expression of key enzymes that are initially limiting to growth ultimately permits population expansion. Mechanistically, increases in expression typically occur either through genetic amplification caused by gene duplications or through loss-of-function mutations that cause deregulation, leading to constitutive expression. Both mechanisms appear to be common in microbial systems adapting to novel carbon sources (Chapter 5).

One effect of population expansion is an increase in the likelihood of persistence in the novel environment. A second is that the effective supply of beneficial mutations, which are the product of population size and mutation rate, is higher. With more mutations available, gene function can be modified and refined by selection, leading eventually to genetic and phenotypic divergence. This model, which has come to be known as the exaptation-amplification-divergence (EAD) model for the evolution of novelty, provides a mechanistic explanation for the evolution of "key innovations," the traits associated with access to novel ecological opportunities.

Two microbial diversification studies illustrate the key elements of the EAD model at work. In *P. fluorescens,* the transition from broth-colonizing *smooth* to biofilm-forming *wrinkly-spreader* involves changes in the regulation of what is known as a two-component signal transduction pathway. Signal transduction pathways in bacteria sense and respond to environmental signals; the best known is the *Che* chemotaxis pathway in *Escherichia coli* that controls the direction of flagellar rotation (Baker, Wolanin, and Stock 2006). The transition to *wrinkly-spreader* often occurs through a range of unique loss-of-function mutations in a gene, *wspF,* that's part of the first component of the two-component system, the *wsp* operon. The effect of mutations in *wspF* is constitutive expression of a signaling molecule, c-di-GMP, produced by the gene *wspR* and activation of the second component, the structural operon (*wss*) that codes for the production of the polymer that forms the biofilm (Bantinaki et al. 2007). A cartoon model of this process was shown in Figure 1.5. To date at least two other receiver operons (*aws* and *mws*) are known that are capable of feeding c-di-GMP signals to *wss*. Mutations in these operons can also lead to *wrinkly-spreader* phenotypes, thus helping to explain the high repeatability of this radiation (McDonald et al. 2009).

The second example comes from the analysis of *E. coli* lines propagated for approximately 1000 generations in mixtures of glucose and acetate as carbon sources (Spencer et al. 2007). *E. coli* typically uses these substrates sequentially, glucose first and then acetate, in a pattern of growth known to microbiologists as diauxie. Diauxic growth is caused by catabolite repression, a form of genetic regulation that prevents the expression of genes permitting use of non-preferred resources (Görke and Stülke 2008). Three of 12 populations diversified into two distinct ecotypes that differed in the timing of their switch to acetate. The ancestral type, known as the slow-switcher (SS), is a superior competitor on glucose but takes longer than the derived type, the fast-switcher, to undergo the diauxic shift onto acetate once glucose is used up. The transition from slow-switcher to fast-switcher occurs in part because the insertion of a transposable element called *IS1* caused a loss-of-function in the isocitrate lyase repressor (*iclR*) gene. *iclR* normally regulates the expression of the acetate operon, *aceBAK,* which is usually repressed during growth on glucose. The mutation caused by *IS1* led to constitutive expression of *aceBAK* and so faster switching to growth on acetate. Interestingly, this single mutation cannot account for all the changes in metabolism associated with the evolution of the fast-switcher, because reverse genetic analysis with *iclR* showed that although switching time was reduced in the genetically modified ancestor, it was not as short as the evolved strains.

These examples involve a common theme: rapid adaptation is facilitated by loss-of-function mutations involved in regulating a key metabolic process—either biofilm production in *P. fluorescens* or the ability to catabolize acetate in *E. coli*—required for invasion of a novel niche, the air–broth interface for *P. fluorescens* and the metabolism of acetate in *E. coli*. Loss-of-function mutations are expected a priori to be more common than gain-of-function mutations, at least at the genetic level, because gene function usually can be disrupted in many more ways than it can be improved. The genetic systems those mutations regulate (*wss* for *P. fluorescens* and *aceBAK* for *E. coli*) are already present in the genome, meaning that they represent exaptations. The mutations themselves led to increased expression of *wss* and *aceBAK,* respectively, allowing the population to invade the novel niche. Moreover, in *P. fluorescens* at least, the stage is set for divergence of gene function: a wide range of sequence changes confer the *wrinkly-spreader* phenotype, and they are associated with a similarly diverse array of fitness values, suggesting the opportunity for selection among independently arisen *wrinkly-spreaders*. Consistent with this prediction, Meyer et al. (2011) have shown that *wrinkly-spreaders* evolve over the course of a radiation to become better adapted to the air–broth niche.

Rapid Evolution of Trade-offs

Divergent selection is very effective at generating trade-offs, a point that was made at length in Chapter 3. What sets adaptive radiations apart from other instances of niche specialization is that trade-offs in a radiation evolve very quickly. For this to happen, adaptation itself must occur quickly. We saw that this was indeed the case in Chapter 2, where the bulk of fitness increases during an adaptive walk occur early in adaptation. The increases often take place within the first hundred to a thousand generations, and they involve a handful (from one to four) of beneficial mutations. The rate of adaptation thus does not represent a major barrier to the progress of adaptive radiation.

Presumably, the trade-offs underlying specialization also evolve quickly. This can happen in two ways. The first is if the direct response to selection in a novel environment exceeds the indirect or correlated response in other environments (Falconer 1990; Kassen 2002). Trade-offs of this sort are commonly observed in microbial experiments involving simple selection in single-substrate environments and in environments containing complex mixtures of resources, as we saw in Chapter 3. This effect probably is seen because the founding genotypes used are not well adapted to laboratory culture. Thus the bulk of genes substituted initially improve adaptation to the general conditions of culture, a situation that may often be analogous with the initial stages of adaptation during adaptive radiation.

The second way that trade-offs arise is from a cost of adaptation, in the sense that the evolved type has lower fitness than the ancestor it was derived from in its original environment. The sources of costs of adaptation—antagonistic pleiotropy and mutation accumulation—were discussed in Chapter 3. In the context of adaptive radiation, antagonistic pleiotropy clearly drives niche specialization in a number of experiments such as the emergence of *wrinkly-spreaders* (Bantinaki et al. 2007) and metabolic specialization (Knight et al. 2006; MacLean, Bell, and Rainey 2004) in *P. fluorescens* and the fast-switchers in *E. coli* (Spencer et al. 2007). Currently no studies are known that have directly implicated mutation accumulation as a significant source of costs during adaptive radiation. This is not surprising, because it would be hard to explain the speed of most adaptive radiations if mutation accumulation were the sole source of trade-offs.

LIMITS AND CONSTRAINTS

One of the most striking features of adaptive radiations is how much they can vary in size, that is, the number of species they contain. Fifteen species of Galápagos finch have evolved from a common mainland ancestor within the last 3 million years, whereas hundreds if not thousands of cichlid species have evolved in a third of that time in the African Rift lakes. The situation is even more striking when considering that replicate radiations, those where the ecological opportunities available to a common ancestral lineage are apparently the same, vary in size. Why should this be so? Various answers have been supplied over the years, all hinging on the rather slippery notion of factors that might constrain or prevent adaptive radiation.

Phylogenetically Distinct Radiations

The simplest and most common explanation for the variance in the size of radiations is variation in the extent of ecological opportunity. The more niche space is available, the more distinct types can be supported. We have already considered many examples where diversification was more extensive in complex compared to uniform environments. What's lacking are more quantitative tests relating niche availability to the extent of diversification.

Genetic constraints may also play a role. As explained in Chapter 7, selection from standing genetic variance can be constrained by negative genetic correlations among traits (Blows and Hoffmann 2005). In microbes the absence of appropriate genetic variance is not typically a major impediment to selection, because the large population sizes used in most experiments ensures a steady supply of new variants through mutation. Nevertheless, the absence of appropriate genetic variance can clearly limit responses to selection in larger organisms with smaller effective population sizes.

Still, the accessibility of beneficial mutations can be a major constraint on adaptive evolution and diversification, even in microbes. The previous chapter pointed out that diversification can be less likely to occur because beneficial mutations are harder to come by as a population nears an adaptive peak (Buckling, Wills, and Colegrave 2003) or because of epistasis (Woods et al. 2011). Jasmin and Kassen (2007a) have further shown how the mutation supply rate can bias the outcome of selection in heterogeneous environments: adaptation in their experiments occurred preferentially toward the substrate with the highest supply rate of beneficial mutations.

A third form of genetic constraint arises due to the stochastic nature of mutation. The physical occurrence or phenotypic expression of a given mutation associated with invasion of a novel environment or niche specialization may be contingent on prior mutations having occurred. Hall et al. (2010), for example, showed that genotypes of *P. aeruginosa* that were isogenic except for the identity of single mutations conferring resistance to the drug rifampicin accumulated different mutations when propagated in a drug-free environment. The implication here is that small genetic differences among founding genotypes can drastically change the genetic routes to adaptation that are available.

More directly relevant to understanding diversification during an adaptive radiation is the evolution of citrate metabolism in the long-term selection lines of *E. coli* founded by Richard Lenski (discussed in Chapter 7). Recall that under the aerobic conditions imposed by the experiment, the ancestor of this experiment cannot metabolize citrate even though citrate is available in the growth medium. Yet, after 31,500 generations, one replicate population evolved the ability to metabolize citrate; this led to the stable coexistence of citrate and glucose specialists, respectively, in this population. Moreover, clones derived from the recent past were more likely than older clones to evolve the ability to metabolize citrate, suggesting that the particular sequence of substitutions occurring before the emergence of a citrate metabolizer was important in allowing this genotype to evolve. We'll consider the genetic details of how this happens in more detail in the next chapter.

Replicate Radiations

Ecological opportunity and genetic constraints cannot, by definition, explain the variance in size among replicate radiations, where the founding population and the ecological opportunities available appear to be identical. What, then, does? Kassen et al. (2004), building on previous work (Buckling et al. 2000; Kassen and Bell 2000), suggested that ecological gradients of productivity (rate of biomass production) and disturbance (rate of biomass removal) can constrain the equilibrium level of diversity achieved in a radiation by decoupling the conditions for evolutionary diversification from those for ecological maintenance of diversity. In their experiment, diversity of *P. fluorescens* in static microcosms was maximal at intermediate levels of productivity and disturbance and declined at the extremes due to the pleiotropic costs of niche specialization. Mechanistically, frequent disturbances and low productivity prevented the formation of a coherent biofilm, which requires expression of an energetically costly cellulose-like polymer. When disturbances are rare and productivity high, by contrast, the mat forms but eventually collapses under its own weight, a process hastened by *smooth* morphotypes that act as cheats in the mat, gaining access to the abundant oxygen at the air–broth interface but contributing nothing to mat strength (Rainey and Rainey 2003). It's important to note that diversification of the *smooth* ancestor into *wrinkly-spreaders* still occurred when disturbance was infrequent and productivity high, as evidenced by the ability of rare *wrinkly-spreaders* to invade a population of *smooths* even though they could not be recovered from the experimental cultures. These results suggest that, despite the presence of strong selection favoring diversification, *wrinkly-spreaders* had a reduced competitive ability that prevented them from being maintained in the population at appreciable frequencies.

The constraints on diversity imposed by gradients of productivity and disturbance can be modulated by parasites (Benmayor et al. 2008) or predators (Hall, Meyer, and Kassen 2008). In the presence of shared enemies, the unimodal diversity patterns become flatter in a way suggesting that diversity is maintained under a broader range of both disturbance rates and productivities. This effect is attributable to a trade-off between competitive ability and enemy resistance: at low productivities, for example, *wrinkly-spreader* is normally outcompeted thanks to the large costs associated with biofilm formation. However, this effect is offset by the advantages associated with the protection from infection or predation afforded by the biofilm itself.

The vagaries of history and chance may also play a role in governing the size of replicate radiations. Fukami et al. (2007) showed that modest changes in the timing and order of immigration drastically changes the course of the *P. fluorescens* radiation in static microcosms. The introduction of a novel *wrinkly-spreader* before or just after inoculation of the ancestral *smooth* genotype removes the ecological opportunity afforded by the air–broth interface and so prevents *smooth* from diversifying.

CONCLUSIONS

The history of life has been punctuated by unusually spectacular periods of evolutionary diversification called adaptive radiation. Understanding the causes of adaptive radiation, and whether they are worthy of being considered unique or somehow special events distinct from more mundane and slower examples of diversification, has been a major focus of evolutionary research.

The ecological theory has been put forward as the leading explanation for adaptive radiations. In Schluter's evaluation of the evidence bearing on the status of the ecological theory he concluded, based on nearly 50 years of comparative and experimental research on adaptive radiations in natural populations of macroorganisms, that the ecological theory "should be regarded as one of the most successful theories of evolution ever advanced" (Schluter 2000a, 242). The work on adaptive radiations in microbial model systems launched by Rainey and Travisano's seminal 1998 paper has produced no substantive reason to question the truth or validity of that statement.

The major results coming from microbial studies of adaptive radiation are as follows:

1. Ecological opportunity is a necessary precondition for adaptive radiation, because it generates divergent selection.

2. Abiotic and biotic factors can generate ecological opportunity, but the former are more important in driving the bulk of diversification in microbial systems.

3. Resource competition is an effective means of accelerating divergence because it can create disruptive selection.

4. There is no evidence that other ecological interactions like predators or parasites contribute to diversification through disruptive selection; more commonly, they do so by creating new ecological opportunities in the form of enemy-free space.

5. The exaptation-amplification-diversification model for the evolution of new gene functions can be used to extend the ecological theory to include genetics.

6. Adaptation to a novel environment often occurs through simple genetic changes involving loss-of-function mutations that deregulate previously inducible pathways or small-scale gene duplications that provide the basic raw material on which selection can then act to generate phenotypic divergence.

7. Rapid diversification and niche specialization often come about through the selection of a few mutations with strongly antagonistically pleiotropic effects.

8. Constraints on adaptive radiation come from any factor that makes accessing beneficial mutations relevant to diversification more difficult or decouples the evolution of diversity from its maintenance; both genetic and ecological factors can act as constraints.

Thus the basic elements of the ecological theory stand the scrutiny of experiments, at least with microbial systems. Diversification happens when a lineage gains access to ecological opportunity and thus finds itself in a regime of strong divergent selection. Most instances of diversification that are genuine adaptive radiations in nature probably also involve an added push to diversification that comes from resource competition. Resource competition, in combination with ecological opportunity, generates strong disruptive selection that drives diversification.

The microbial approach has allowed scientists to extend and expand the ecological theory in ways that haven't, until recently, been possible with larger organisms, most importantly by incorporating genetics. Here, the major advance has been to recognize, first, that a characteristic sequence of events are involved in the initial stages of diversification in novel environments. These events are captured in the EAD model and so provide a mechanistic explanation for adaptive peak shifts. Peak shifts occur through rather simple genetic changes—often loss-of-function mutations that deregulate previously inducible pathways or small-scale gene duplications—that provide the basic raw material that selection can then act on to generate phenotypic divergence. So the crossing of an adaptive valley does not require wholesale reorganization of the genome but, instead, fairly simple and common genetic changes that result in dramatic gains of function at the level of the phenotype that matters most, fitness.

A second, equally important insight has been that the rapid adaptation and diversification into niche specialist lineages that is the hallmark of adaptive radiation is consistent with models of adaptation involving beneficial mutations with variable effect sizes. These models were discussed in detail in Chapter 2, so we won't go into them again here except to note that they predict the bulk of adaptation to be accomplished through the substitution of just a few beneficial mutations. The evidence from microbial systems suggests that it can be as few as one mutation and not usually much more than a handful. When these mutations both increase fitness in the selection environment and decrease it in other environments, the evolution of niche specialization can happen very quickly. Most adaptive radiations therefore probably involve the selection of mutations with niche-specific effects.

Integrating genetics into the ecological theory can help researchers understand more precisely the conditions under which adaptive radiations do, or do not, proceed. Perhaps most telling in this regard is the observation mentioned in the preceding paragraph, that adaptation to novel environments does not impose a major barrier to diversification in asexual systems. If this is also true in sexual systems, then the main obstacle to adaptive radiation is the speciation process. Adaptive radiations in sexual systems will be most apparent when adaptive diversification in niche space is tied to reproductive isolation, as in models of ecological speciation. Another example is the notion of constraints on adaptive radiation. This idea has been around for some time, but specifying precisely what those constraints are and how they act to limit diversification has proved challenging. Microbial systems provide clear evidence for genetic constraints imposed by the mutational landscape, the spectrum of single-step mutations that contribute to adaptation and diversification, ecological constraints imposed either by the extent of ecological opportunity itself or by ecological gradients of productivity or disturbance, and stochastic constraints associated with the timing of invasion of a novel niche. Recognizing the importance of these factors, as well as the many others not explicitly studied in microbial systems—such as those imposed by habitat loss or sexual selection, to name but two—can take us a long way toward explaining why and how adaptive radiations happen in the first place, and why when they do they can vary so dramatically in size.

Some compelling and important issues remain, and microbial systems are uniquely positioned to help answer them. Three issues stand out as major gaps in our knowledge.

The first is the long-term fate of diversity. Do radiations reach a sort of carrying capacity after the initial burst of diversification, or do they continue to diversify? These issues were discussed in the broader context of models of diversification in general in Chapter 7, and we saw a number of examples of saturating diversity dynamics. In reality, many radiations actually lose diversity in the long run, albeit more slowly than they gained it during the initial phases of the radiation, after the bulk of ecological opportunities are filled. This so-called boom–bust dynamic is seen in the *P. fluorescens* radiation (Meyer et al. 2011). Schluter (2000a) has suggested that it involves continued strong disruptive selection leading to the success of more phenotypically extreme types and the loss of intermediate ones. Experimental results are consistent with this idea: the two main niche specialists, *wrinkly-spreader* and *smooth,* coexist at all times in the radiation, implying strong disruptive selection. Moreover, the loss of diversity is associated primarily with the loss of variant *wrinkly-spreader* morphotypes, and this is due to continued adaptation within the *wrinkly-spreader* niche. Notably, the metabolic profile—a measure of resource-use phenotype, becomes increasingly divergent as the radiation ages, in line with Schluter's suggested mechanism.

The speed and predictability of adaptive radiations present another major unresolved issue. Clearly, adaptive radiations can occur extremely rapidly and repeatedly when few genetic changes that have strongly antagonistic effects are responsible, as in the *P. fluorescens* radiation. But the extent to which this system is indicative of most radiations, in either microbes or macrobes, remains unresolved. An answer can come only from the detailed genetic analysis of adaptation in more systems. The key issues here are to identify the specific molecular targets of adaptation, their effects on fitness, and how many of them are involved in adaptive diversification. Whole genome sequencing will likely be the most effective, albeit brute-force, approach to this task. Yet it stands to shed important light on the range of genomic changes involved in adaptive diversification.

Lastly, the relative importance of standing genetic variation as opposed to mutation in generating adaptive divergence also remains under-studied. All the microbial experiments reviewed here focus exclusively on variation arising de novo through mutation. Although standing genetic variation no doubt plays an important role as well, whether it constrains adaptive divergence, and if so for how long, remains unclear. Microbes provide a relatively easy way to get an answer. Genetically diverse founding populations can be easily constructed from crosses between strains of facultatively sexual species such as yeast or unicellular algae. One could then contrast the rate and extent of diversification in experimental lines that varied in the quantity of standing genetic variance. A further advantage of using facultatively sexual species to study diversification is that it's an opportunity to tie the evolution of niche specialization directly to the evolution of reproductive isolation. The extent of correlation between these two processes, as noted previously, is key to understanding how adaptive radiation proceeds in sexual systems.

Let's end this chapter by returning to the issue of how important adaptive radiation has been for creating life's diversity. The consensus seems to be that radiations are the major way that diversification happens, but this view is at odds with their apparent rarity in nature. The word *apparent* applies here, because we don't yet know whether adaptive radiations really are rare. Have we simply missed many radiations because we notice only the most spectacular ones, or only those that occur in unusual conditions like islands or lakes? Microbial experiments give us a clear picture of the different ingredients required for a radiation to occur: ecological opportunity in combination with resource competition and, to boot, abundant and appropriate genetic variation. Too many factors—and here's where the various genetic and ecological constraints that can derail an incipient radiation come in—need to be in place for a lineage to undergo the spectacular, some even say exuberant, diversification of form and habit that characterize adaptive radiations. So it's tempting to say that an unusual alignment of forces is required for a radiation to proceed, and this makes adaptive radiation a genuinely rare occurrence. On the other hand, diversification may be a process that proceeds at variable rates and to different extents even if not all the ingredients are in place to produce an adaptive radiation per se. Under this view, adaptive radiations are the extreme tails of a distribution of diversification events that vary in both speed and extent. They are, in other words, one end of a continuum of a more general process of adaptive diversification. Which of these interpretations is a more appropriate description of adaptive radiation remains an important avenue for future research.

CHAPTER 9

Genetics and Genomics of Diversification

ॐWe are all familiar with the good things that microbes do for us, even if we sometimes forget that they are responsible. Fermentation, the process by which microbes convert carbohydrates into alcohols or acids, is one of the most important. Cheese, yogurt, tofu, wine, beer, kimchi, sauerkraut. Every culture around the world has its favorite fermented food. These foods are nutritious, safe, and essential components of the cultural tapestry of our communities. Bravo microbes.

In Africa, many fermented foods are produced using traditional methods and sold in local markets. The inner workings of fermentation—which microbes are responsible and the biochemistry that happens—are pretty much still a mystery for all but a few products. This is unfortunate. A better understanding of fermentation in these traditional products can improve indigenous knowledge and practices and contribute to improving local food security. It may also represent an opportunity for jobs, training, and economic development if the production process can be industrialized.

Besides these very practical uses for African fermented food products, there's a more scientific motivation for studying them. Many products are produced in a way that closely resembles the batch-culture techniques employed in lots of the experiments discussed in this book. Workers start the fermentation process by gathering raw materials such as milk or cereals and sometimes other ingredients like the roots of local plants, which are thought to provide key enzymes that help the fermentation along. These are mixed with a small amount of a previously made product, often in a previously used culture vessel. The culture vessel is typically a large (up to 40 L) squash gourd called a calabash, shown in **Figure 9.1a**. The insides of calabashes are never washed and so develop a biofilm that, along with the small amount of the old product, contains the seed community for the fermentation. The culture is left for a few days to ferment and the finished product is then collected and sold,

Figure 9.1 Fermented food products from Zambia. (a) A calabash used for fermentation. (b) A finished milk-based product, called mabisi, being served in a market.

save for another small amount that's used to seed the next batch. In this way, a community of microbes evolves together for prolonged periods. The products are safe, nutritious, and tasty, three features that can be used to describe the "function" of this microbial ecosystem. Traditionally produced fermented foods thus provide an opportunity to study the evolution of an entire, often diverse, community of microbes with known ecological function.

I was introduced to fermented products from Zambia by my former postdoc, Sijmen Schoustra, who lived for three years in Lusaka while teaching at the University of Zambia. Sijmen has spent the last couple of years investigating the microbial ecology and biogeography of three fermented products, called munkoyo, chibwantu, and mabisi (pictured in **Figure 9.1b**), produced in Zambia. The first two are made from cereals including maize, millet, and sorghum; the last is made from milk. They are nonalcoholic beverages that have a slightly sour taste. Many people consume them daily as a refreshing and nutritious lunch and occasionally as part of traditional ceremonies.

These products are especially attractive for understanding the link between diversity and ecosystem function, for two reasons. One is that producers do not share their cultures with each other. In other words, the microbes do not migrate among producers, and each product that's sold in a market has evolved independently from all others. We don't yet know how long they have been evolving—it might be hundreds, thousands, or millions of microbial generations. But we suspect, based on estimates of the volumes used and cell density counts in finished products, that to initiate a new culture requires a similar number of doublings as a typical microbial experiment in the lab, meaning anywhere from six to 10 generations at least. The second reason microbiologists are interested in these beverages is that lots of producers are selling the same product in different locations around the country. This means that we can ask questions about how entire communities evolve, and we can also ask how much variation there is from seller to seller in the constituent microbial communities themselves (Schoustra et al. 2013).

What does all this have to do with the genetics and genomics of diversification? In the first place, DNA sequencing techniques are essential for finding out what microbial species actually are present in these cultures. Conventional PCR-based sequencing revealed the products are dominated by acid-producing bacteria like *Lactobacillus*. These species lower the pH and presumably give them their slightly sour taste. The low pH probably also helps protect the products from unwanted pathogens. Lots of variation is also found in microbial community composition, both among the different products themselves and from seller to seller of the same product. Two samples of chibwantu from different sellers in the same market, for example, had completely different microbial profiles: one was dominated by *Weisella* species, bacteria that produce lactic acid and are often found in kimchi, and the other by *Lactobacillus*. Clearly, very different sets of microbes were serving the same ecological function.

In the second place, many sources of divergent selection must be at work in these communities. The recipes and ingredients for each product are different, for example, and so constitute one cause of selection favoring some genotypes or species over others. The biotic environment within each fermentation vessel is likely to differ as well because the products are made at home, not in the controlled conditions of a laboratory. Additionally, the identity and abundances of whatever species do end up in a given fermentation will change over time as the community evolves through repeated rounds of subculture and growth. Even the sellers themselves might represent a source of selection that generates diversity. When interviewed, sellers say they try to achieve slightly different tastes from each other if they are selling the same product in the same market. This could explain why the chibwantu samples mentioned earlier contained very distinct microbial communities: perhaps the different species give the product slightly different flavors. It seems that by competing for customers, sellers are actually driving diversification in the microbial community.

As with more familiar instances of lineage diversification and community assembly, for example adaptive radiation and the colonization of islands, uncovering the forces governing diversity in the microbial communities of fermented food products will be a prodigious task. Working with microbes makes the task a little easier, because the experiments don't take quite as long as they do in, say, birds or spiders. Microbiologists face other challenges, however; for one thing, many of the species we identify in the products cannot be easily cultured in the laboratory. Nevertheless, Sijmen has found that it's possible to make some of the products, or at least a close copy of them, in the lab by using store-bought ingredients. So there's a real possibility of understanding the very practical question of how these products are made and what makes them safe. We also may be able to get deeper into the issue of how they become so diverse and so different, at the microbial level, from each other.

For the moment, we have what almost everyone else has when they begin to study diversification: lists of species in different environments. The environments here are the different samples of product. We know that the microbial communities differ among products as well as within different samples of the same product. We would like to know more about how that diversification occurred. Ecological conditions are clearly important. The culture vessel is no doubt a highly heterogeneous environment, akin perhaps to static microcosms of the *Pseudomonas fluorescens* radiation in the diversity of nutrients available and the concentration gradients that develop. The community of microbes that exist in the environment as well as on the biofilm that forms inside the culture vessel is also important because it's the community that seeds the fermentation. But characteristic genetic changes must also be evolving, so that some genotypes can coexist within this highly heterogeneous environment and, at the same time, fermentation can take place. Knowing about these genetic changes is thus integral to understanding how these products are made and how their ecological functions—taste, safety, durability—evolve.

The example of Zambian fermented products represents another, more fundamental gap in our understanding of the process of diversification. In general, evolutionary biology has focused almost entirely on the ecology of diversification and ignored the genetics underlying it. Our ignorance of the genetics of diversification stems in part from the very sensible argument that if we don't first understand what ecological conditions promote diversification, knowing the genes involved won't help much. It also comes from the fact that, until recently, we haven't been able to find the genes that matter without a tremendous amount of effort and no small amount of luck.

Things are changing. As emphasized in previous chapters, biologists have a better understanding of the ecological conditions that promote diversification. Ecological opportunity and the divergent selection it creates, helped along by ecological interactions such as resource competition, are effective ways of generating diversity. With the introduction of relatively low-cost next-generation sequencing, it's now possible to scan entire genomes for sites showing signatures of strong divergent selection. The information from such genome-wide surveys can be used to supplement and extend ecological theories of diversification. In other words, a truly general theory of diversification, one that integrates ecology and genetics, will soon be within reach. This chapter provides a first sketch of the genetic side of a theory of diversity.

Is it possible to say anything general about the genetic changes underlying diversification? Are there, for example, genes for diversification or even speciation? Perhaps this way of asking the question is too strong. We might soften it by asking if there's something more we can say about the nature of the genes associated with diversification that goes beyond what we find when we identify the genes involved in conventional adaptation to a given environment.

In some senses, this question can already be answered with a tentative yes. The theory of ecological speciation, for example, sees reproductive isolation as the inevitable outcome of adaptation to distinct optima. Genes associated with divergence are thus expected to be those that cause adaptation and generate trade-offs in fitness across environments (Nosil and Feder 2012; Nosil, Funk, and Ortiz-Barrientos 2009). Others have suggested that so-called speciation genes, those that decrease hybrid fitness between incipiently divergent populations, are responsible for diversification independently of ecological divergence (Presgraves 2010). Developmental biologists, on the other hand, have emphasized that major changes in morphology often seem to be associated with mutations affecting gene regulation during development rather than those influencing the structural or functional properties of enzymes (Carroll, Grenier, and Weatherbee 2001; Stern and Orgogozo 2009).

Making sense of these confusing and at times contradictory claims about the genes or genetic architectures associated with divergence remains a real challenge. What's needed is a more comprehensive approach that identifies the genetic changes accompanying divergence and quantifies their contribution to phenotypic disparity and reproductive isolation. Microbial experiments offer an outstanding opportunity in this regard. Unfortunately, at the time of this writing, preciously few studies taking on this task are under way or already published. A synthetic treatment of the genomics of diversification along the lines of that presented in Chapter 5 for adaptation is therefore still some way off.

Instead, we have a small number of case studies where diversification was observed in a microbial experiment and some attempt was made to identify the genetic changes responsible. Some have taken a candidate-gene approach to identifying the loci associated with diversification or have examined patterns of gene expression rather than sequence changes. Few studies, however, beyond a handful in phage, have made whole genome sequencing data available for lines that have been selected in different environments.

THE GENOMICS OF DIVERGENT SELECTION

Let's begin by considering the results of experiments that have examined whole genome sequences following selection in divergent environments. How many genetic differences typically distinguish a pair of lineages that have evolved in different environments? Are the genetic changes that evolve distributed evenly throughout the genome, or are they clustered in particular regions? What sort of genetic changes, if any, are associated with reproductive isolation? We'll then discuss the genetic architecture of diversification and use this as a first sketch of a genetic theory of diversification that builds on and extends the exaptation-amplification-diversification (EAD) model introduced in Chapter 8.

Parallel Evolution Within and Between Environments

Divergent selection is a crucial first step in ecological diversification because it leads to niche specialization (Chapter 3). To what extent do divergent niche specialists differ at the genetic level? To find out, we can look at data taken from the literature on the number and identity of genomic substitutions that occur when the same founding genotype is selected in different environments. From whole genome sequence data reported in the papers, the numbers of shared and unique mutations substituted between all pairs of independently evolved lines from the same founding genotype were counted. This data was then used to estimate the degree of parallel genomic evolution between lines in two ways: as the number of unique mutations substituted every 100 generations and as the Jaccard similarity coefficient, which measures the probability that a given mutation is shared by any pair of lines. Pairs of lines evolved in the same environment should show more parallel evolution than those from different environments.

The results all come from phage experiments and are summarized in **Table 9.1**. Compared to selection in the same environment, divergent selection leads to an average of about 1.5 more unique substitutions per 100 generations compared to selection in the same environment (**Figure 9.2a**). The extent of parallel evolution measured using the Jaccard coefficient is lower following divergent selection compared to uniform selection, although not always significantly so. (A brief technical note here: The probabilities that Jaccard coefficients are the same for uniform and divergent selection reported in the table are calculated by randomizing the observed mutations among selection lines 10,000 times and recalculating the Jaccard coefficient each time. This approach is appropriate because the Jaccard coefficient involves pairwise estimates of similarity among lines and so precludes the use of conventional parametric statistics.) Combining probabilities using Fisher's method gives a highly significant result, however ($\chi^2 = 63.9$, $df = 14$, $P < .0001$), suggesting that parallel evolution really is less likely following divergent selection as compared to uniform selection.

Although these results are all based on work with phage, it's worth mentioning that researchers have seen similar results in bacterial experiments. Typically, parallel evolution is less likely between treatments than within treatments.

Table 9.1 Parallel Genomic Evolution in Divergent Environments

Study organism	Genome size (bp)	Gene number	Reference	Selection environments	Generations	Total Unique Changes		Correlation		Jaccard Similarity		Probability
						Within	Between	Within	Between	Within	Between	
phiX174	5386	11	Bull et al. 1997	high temperature on either *E. coli* or *S. typhimurium* host cells	1000	15	20.3	0.24	-0.07	0.28	0.12	<0.0001
S13	5400	11	Wichman et al. 2000	high temperature on either *E. coli* or *S. typhimurium* host cells	1000	14	15.4	0.01	-0.08	0.14	0.11	0.184
phiX174	5386	11			1000	14.3	20.8	0.16	-0.23	0.24	0.05	<0.0001
VSV	11,200	5	Novella et al. 2004	hamster and sandfly host cells	240	3.25	5.17	0.38	-0.09	0.42	0.06	0.225
VSV	11,200	5	Remold, Rambaut, and Turner 2008	HeLa and MDCK host cells	95	7.08	10	0.25	-0.1	0.3	0.08	0.001
DENV/IQT	10,713	14	Vasilakis et al. 2009	HUH and C6/36 host cells	30	1.5	3.5	0.42	-0.51	0.62	0	0.153
DENV/P8	10,719	14	Vasilakis et al. 2009	HUH and C6/36 host cells	30	4.5	5.5	-0.11	-0.38	0.1	0	0.21
Grand mean								0.19	-0.21	0.3	0.06	0.06

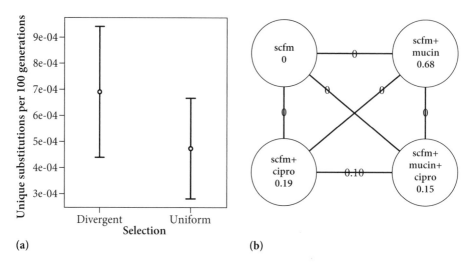

(a)   (b)

Figure 9.2 Genomic differentiation and parallel evolution as a result of divergent selection. (a) Substitution rates (per 1000 generations) of unique nucleotides following divergent and uniform selection in laboratory experiments with phage. (b) Parallel evolution, measured as the Jaccard similarity index, is more likely among independently evolving lines from the same environment (circles) than between environments (edges) in evolving populations of *Pseudomonas aeruginosa.* Note: scfm (synthetic cystic fibrosis medium); cipro (ciprofloxacin). Figure reproduced with permission from Wong, Rodrigue, and Kassen (2012).

A representative result is shown in **Figure 9.2b** from a recently published study where experimenters selected *Pseudomonas aeruginosa* for approximately 50 generations in environments that resembled, to different degrees, the lungs of cystic fibrosis patients with and without treatment with fluoroquinolone antibiotics (Wong, Rodrigue, and Kassen 2012). More work on the subject of parallel evolution under different modes of selection would be welcome.

Niche Dimensionality

In reviewing experiments that attempted to generate speciation in the laboratory, Rice and Hostert (1993) pointed out that multidimensional or multifarious divergent selection—selection caused by adaptation to multiple niche dimensions or imposed by selection on multiple traits—coupled with reduced gene flow was the most effective means of achieving reproductive isolation. This result may have occurred because the strength of selection on a particular genomic region was stronger, or because more genes are involved in responding to selection in multiple directions. Either way, it has often been suggested that phenotypic differentiation and reproductive isolation evolve more readily when divergent selection is multifarious, acting on multiple niche dimensions or traits at the same time. Consistent with this idea, Nosil and Sandoval (2008) showed that the reproductive isolation that occurs among species pairs of phytophagous walking stick insects (*Timema*) on different host plants involves divergence in both cryptic coloration and physiology. But among ecotypes on the same host plants within a species, they found that only cryptic coloration is divergent.

Whether the genetic causes of divergent selection under multifarious selection are due to a few genes of large fitness effect, or many more of smaller effect, remains unknown in the *Timema* system or any other, for that matter. This question has been investigated in experiments that observed the opportunistic pathogen *P. aeruginosa* as it evolved in environments that closely resembled the lungs of people with cystic fibrosis (CF). They selected a single, ancestral strain under different combinations of stressors. The stressors used were mucin—basically snot—designed to mimic the highly viscous nature of the CF lung, and the commonly used antibiotic ciprofloxacin. Whole genome sequencing of evolved lines revealed mutations that had phenotypic effects on both motility (which tends to be lost when the bugs form microcolonies and biofilms in the lung) and resistance similar to those seen in "real" infections. Over the course of the 50 generations of selection, the lines selected with a single stressor, either mucin or ciprofloxacin, fixed on average one mutation. When both stressors were present, though, the researchers found two mutations fixed, on average. This result thus supports the idea that the stronger selection imposed by multifarious selection stems from the multiple mutations required for adapting to these more complex environments.

Genomic Clusters of Divergence

Next-generation sequencing techniques make it possible to scan large regions of the genome in non-model organisms for signals of genetic differentiation that exceeds what's expected through neutral processes. The basic procedure is to collect samples of individuals from different populations within a species or among closely related species, sequence as much of their genomes as possible, and then look for regions of unusual genetic differentiation. Because recombination results in sharing of genes among genomes, only those regions under strong selection, or those closely linked to them, will become highly differentiated. Such "genomic islands of divergence" can be quantified using various measures of allelic differentiation such as the fixation index (F_{st}). Genomic regions under strong selection are identified as outliers in a scan across the genome for whatever metric of differentiation is used. This work represents significant methodological challenges because loci under selection need to be confidently distinguished from false positives that are not genuine outliers. In addition, different demographic models can give different answers, complicating matters even further (Barrett and Hoekstra 2011).

Technical challenges notwithstanding, in natural populations the total amount of genome under strong divergent selection seems to be quite small. A recent review estimated the fraction of the genome affected by divergent selection to be on the order of 5–10 percent (Nosil, Funk, and Oritz-Barrientos 2009; see also Turner, Hahn, and Nuzhdin 2005). Regions of divergence are distributed across the genome rather than being clustered in one or a few spots. Similar results are seen in natural microbial populations that undergo recombination (Shapiro et al. 2012).

In laboratory experiments that have tracked the genomic consequences of divergent selection, the results are not too dissimilar. If we focus on those studies reported in Table 9.1, where there is evidence that parallel evolution is significantly lower between environments than within environments, we can conservatively estimate the amount of genomic divergence by count-ing the number of sites where a mutation has occurred in all replicates from one environment but not the other. Such strongly divergent sites account for between 7 and 8 percent of all variable sites reported. They are also all non-synonymous, providing further evidence that they are under strong selection. Interestingly, the divergent sites are distributed across the genome rather than being clustered in any particular region.

Thus the data from both natural isolates and laboratory experiments, limited though it is, seems to be telling us pretty much the same story. The genomic footprint of divergent selection is recorded in strong divergence at just a few sites spread around the genome. This is true for natural popula-tions where recombination occurs as well as laboratory ones where it doesn't, suggesting that divergent selection in nature may often be strong relative to recombination.

Genetic Targets of Divergence

For all but a small number of examples—coat color in mice and body armor in sticklebacks are among the best known—we know very little about the genes and genetic systems responsible for ecological divergence. It's no mystery why: it takes a lot of work, and no small amount of luck, to find the genes in the first place and then to characterize the nature of selection on them (reviewed in Barrett and Hoekstra 2011). It would seem reasonable, then, to turn to microbes for some help because at least here we stand a chance of identify-ing mutations that have been selected and quantifying their contributions to divergence by estimating their fitness effects across environments. Unfortu-nately, with one exception, these experiments have not yet been published.

The exception is a study by Anderson et al. (2010), who identified the genetic targets of divergent adaptation in yeast populations that had evolved in either high-salt or low-glucose environments for 500 generations (Dettman et al. 2007). Adaptation to high salt involved mutations in genes required for maintaining an appropriate osmotic balance, either by adjusting proton (*PMA1*) or sodium ion efflux (*ENA, CYC8*). Adaptation to low glucose came through mutations in two genes involved in different phases of the growth cycle, one affecting exponential phase (*MDS3*) and the other stationary (*MKT1*) phase. These results are interesting for what they tell us about how yeast responds to these different kinds of stress. It would also have been nice to know what the fitness effects of these alleles were in the environment that they did not evolve in, but unfortunately those experiments were not done. However, the researchers did create a double mutant containing the *PMA1* allele that evolved in high salt and the *MKT1* allele that evolved in low glucose and tested its fitness in low glucose. The hybrid had lower fitness than the ancestor, which lacked both evolved alleles. This is an example of a Dobhzansky-Muller incompatibility between *PMA1* and *MTK1*: selection independently fixed these mutations in different environments, and by chance the double mutant turned out to have low fitness. Reproductive isolation has thus begun to occur in these lines due to divergent adaptation to different environments.

Some insight into the genetic targets of divergence can also come from taking a closer look at the phage experiments in Table 9.1. The genes that show the most divergence are often involved in host recognition, as might be expected because in phage much divergent selection stems from the need to infect and replicate in different kinds of host cells. There are mutations in other genes as well, including those involved in replication initiation, RNA polymerase, or host-cell lysis. The fitness effects of these mutations across different host environments remain unknown at this time. It's therefore difficult to say much about how important they are in driving host-specific adaptation and fitness trade-offs.

GENETIC ARCHITECTURE OF DIVERSIFICATION

Certain lineages seem inclined to diversify more extensively or more rapidly than others. Everyone has a favorite example, but here we'll consider Nearctic teleost fishes. These include sticklebacks, whitefish, and salmonids that have repeatedly diversified into ecologically and morphologically distinct species following invasion of low-lying freshwater lakes and streams as the glaciers receded about 15,000 years ago. When these lineages are contrasted against the remarkable evolutionary constancy of so-called living fossils like horseshoe crabs or gingko trees, it's easy to wonder whether they might have some special quality that makes them prone (or not) to diversify.

The ecological context in which a lineage finds itself clearly plays a role in governing whether it diversifies, as we saw in Chapters 7 and 8. In the presence of abundant ecological opportunity and strong resource competition, for example, diversification can be rapid and extensive. However, it's not always obvious how a lineage gains access to abundant ecological opportunity in the first place. Chance may play a role. Two to 3 million years ago, a small population of seed-eating finches from South America were compelled to leave the mainland. They found their way to the Galápagos Islands, 900 km away, where they encountered a wide range of unexploited niches and subsequently diversified into 14 species of Darwin's finches (Grant and Grant 2003). It has also been suggested that particular features of the genetic architecture of a lineage enable it to gain access to the ecological opportunities necessary to drive diversification in the first place. Key innovations like the evolution of phytophagy in insects or endothermy in dinosaurs (Chapter 7), for example, may allow a lineage to gain access to novel ecological opportunities and so promote diversification. Thus certain kinds of genes or genomes may make a lineage more or less likely to diversify. This section discusses some of the more common explanations that have been offered and what microbial experiments have told us about them.

Genetic Mechanisms Promoting Diversification

Various suggestions and guesses have been made about genetic mechanisms that might promote or hinder diversification. One commonly cited diversity-promoting mechanism is *cis*-regulatory mutations, whose effects are expected to be less pleiotropic, and so more likely to be beneficial, than mutations that occur in genes coding regions (Carroll, Grenier, and Weatherbee 2001; Hoekstra and Coyne 2007; Stern and Orgogozo 2009). We saw in Chapter 5 that although *cis*-regulatory mutations may have been important in generating morphological diversity in some metazoans, they play a minor role in diversification of microbial systems. Other suggestions include a range of mechanisms for creating new genes such as exon shuffling, gene translocation, mobile element insertion, gene duplication, and lateral gene transfer (Long 2001). The first three mechanisms, exon shuffling, gene translocation, and transposable elements, seem to be most prevalent in eukaryotes. The last, lateral gene transfer, is largely restricted to prokaryotes. Gene duplication stands out as the only general mechanism that occurs in both eukaryotes and prokaryotes, and there is good evidence that it has been an important source of new genetic functions in a wide range of organisms (Conant and Wolfe 2008; Dujon 2010; Mortlock 1983; Otto 2007). Examples of the importance of gene duplication in adaptation were discussed in Chapter 5. We'll return to this topic in more detail soon.

Genetic Mechanisms Preventing Diversification

Other mechanisms can constrain or prevent diversification. We touched on this idea briefly in previous chapters when looking at genetic constraints on diversification (Chapters 7 and 8). Any factor that prevents a lineage from accessing the genetic variation necessary for adaptive diversification can act as a constraint. Proximity to a peak in an adaptive landscape, epistasis, and the stochastic nature of mutation are examples of genetic constraints that have been documented in microbial systems; genetic correlations that act to prevent a response to selection in a particular phenotypic direction can be important in systems where standing genetic variance is abundant.

The Genetics of Key Innovations and the Evolution of Novelty

The evolution of a key innovation, and of novelty in general, involves a paradox. How does something new come about if all selection has to work with is something old? The answer, of course, is that novelties are not produced from scratch. Evolution is, as François Jacob has said, a tinkerer using the materials already at hand in new ways (Jacob 1977). The various mechanisms put forward to explain the evolution of new genes, for example, all attempt to explain how existing genetic material can be reused in novel ways to produce new functions.

The evolution of novel substrate use in bacteria lends some insight into how novel traits are selected. One recent and particularly striking example of the evolution of a novel trait is the evolution of citrate use in aerobically grown *Escherichia coli* populations (mentioned briefly in Chapter 8; Blount, Borland, and Lenski 2008). The population in which citrate use evolved had been growing in a minimal medium containing glucose as the sole carbon substrate but with citrate added as a chelating agent. The ancestor uses glucose when growing aerobically; indeed, the inability to use citrate under aerobic conditions is a diagnostic trait of *E. coli*. Yet, after 31,500 generations, the population in question evolved the ability to metabolize citrate (Cit+), and this led to the stable coexistence of citrate specialists and glucose specialists. Whole genome sequencing revealed that the evolution of the Cit+ variant came about in four steps (Blount et al. 2012).

The first step was potentiation, the evolution of a genetic background that makes it possible for the lineage to access mutations that make the new function possible. Although the identity of these mutations remains unknown, they clearly made it more likely for the Cit+ phenotype to arise: evolved genotypes isolated closer to the time when the original Cit+ genotype arose during the long-term experiment had higher mutation rates to Cit+ than did the ancestral strain used to found the experiment. Thus the emergence of the Cit+ phenotype is contingent on these previous mutations occurring, although precisely what the selective advantage of these potentiating mutations is, and how they make Cit+ mutations more accessible, remains obscure. The second step was exaptation, whereby a preexisting gene function becomes co-opted for growth and reproduction under novel conditions. Citrate can be used for metabolism once it enters the cell, but the challenge under aerobic conditions is getting it inside the cell in the first place. The Cit+ genotype accomplished this by using a citrate/succinate antiporter (a protein that transports sugars by, in this case, exchanging citrate for succinate) that normally transports citrate under anaerobic conditions. So the origin of aerobic citrate use didn't come from nowhere; it came from an existing function that was pressed into service in a different ecological context. The third step improved the ability to metabolize citrate, which in this lineage came about through amplification of the key genes necessary for getting citrate into the cell under aerobic conditions. A nearly 3000-base-pair (bp) portion of the citrate fermentation operon that includes the gene *citT* was duplicated, and it is *citT* that codes for the transporter that gets citrate into the cell during fermentation. In the ancestor, this gene is not expressed in the presence of oxygen but in the Cit+ variant, the duplicated gene had been inserted just downstream of the promoter of another gene, *rnk,* that directs transcription when oxygen is available. The effect of this rearrangement was to express *citT* and so allow growth, albeit very weakly, on citrate. The fourth step was further growth improvement on citrate, which seems to have occurred, in the earliest stages following the emergence of weak citrate growth, by further amplification of this modified promoter-transporter gene combination. These later-stage improvements have led to increases in population sizes of the Cit+ phenotype and prolonged coexistence with the ancestral, Cit−, phenotype.

This example may be unusual, because the phenotypic innovation—the ability to metabolize citrate, a substrate that by all accounts could not otherwise be metabolized—was so extreme and took so long to evolve. Nonetheless, in many other microbial studies the ability to use what was previously thought to be an unusable substrate has evolved through selection, sometimes intentionally and sometimes not (Hall 1989; Lin, Hacking, and Aguilar 1976; MacLean and Bell 2002; Melnyk and Kassen 2011; Mortlock 1983). In these studies, the time required for the evolution of the novel function is never as long as in Blount's experiments. We find, for instance, that our lab strain of *P. fluorescens* SBW25 repeatedly evolves to metabolize xylose, a substrate that it can barely get by on because it lacks a key gene, *xylB,* in the xylose utilization operon, within a few hundred generations. In these lines, mutations in genes associated with the transcriptional regulator, GntR, are fairly common, suggesting that changes to gene regulation, perhaps through amplification of a gene product that affords some crude ability to metabolize xylose, are the cause of adaptation.

All these cases of the evolution of novelty reflect a common theme that should be familiar from Chapter 8: exaptation that makes use of existing genetic tools for a new purpose, amplification of key enzymes required for metabolism and population growth, and eventually further refinement and divergence by selection to improve function. This is the EAD model that was introduced in the context of adaptive radiation. In light of the Cit+ story, it seems that a more general theory for the genetics of diversification needs to account for variation in how long it takes for novelty to evolve, something that for Cit+ is attributable to the potentiation step. Potentiation may seem rather idiosyncratic, but there is good evidence that it occurs in at least one other case as well, which we discuss as part of the next section.

A GENETIC THEORY OF DIVERSIFICATION: THE PEAD(R) MODEL

A genetic theory of diversification needs to incorporate potentiation as an added step that precedes exaptation in the EAD model. But another missing feature could be usefully added to a genetic theory, one that applies to sexually reproducing lineages: reproductive barriers causing speciation. Here we consider an expanded genetic model for diversification called PEAD(R), for potentiation, exaptation, amplification, divergence, and reproductive isolation (**Figure 9.3**). The parentheses around the *R,* standing for "reproductive isolation," are a reminder that reproductive isolation applies only to lineages that reproduce sexually.

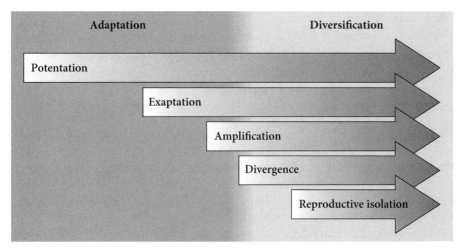

Figure 9.3 The PEAD(R) model for the genetics of diversification. Adaptation to a novel environment involves potentiation, exaptation, and amplification. Divergence results from divergent selection leading to ecological specialization and, in sexuals, reproductive isolation. Details of each step are given in the text. Longer arrows indicate longer times to diversification for a lineage starting from a given genetic position (not to scale).

Potentiation, as already mentioned, is the evolution of a genetic background that affords a lineage access to mutations that would otherwise be inaccessible. A potentiated genotype can then access mutations that confer a novel function. It is not, to be clear, the novel function itself. Rather, potentiation represents an example of how adaptation to one set of conditions brings a lineage into a new region of mutational "space" that allows a novel function to eventually evolve. This seems at first blush to be a highly unlikely event, because the new function evolves only as a by-product of adaptation to other conditions. Nevertheless, the Cit+ story is fairly convincing evidence that it can happen. But if you are not convinced, here is another, also from *E. coli.*

Meyer et al. (2012) investigated how the lytic bacteriophage λ, which is known to infect *E. coli* only via a single host-cell receptor on the outer membrane of the bacterial cell called LamB, might be able to evolve to infect its host through a novel receptor, OmpF. The researchers suspected this could happen because it had been noted that *E. coli* became resistant to λ through mutations in *malT,* a gene that acts as a transcription factor promoting *lamB* expression, but that phage persisted in these populations at low densities. Selection should thus favor any phage mutant that could infect *E. coli* via an alternative route. Subsequent evolution experiments with the *malT*-resistant host and the original phage produced a phage mutant that could infect *E. coli* via the OmpF receptor. Whole genome sequencing revealed that the ability to infect through this novel target required the substitution of four mutations that improved function on the original LamB receptor. Once these mutations had been substituted, the transition to the novel receptor could occur in just a single mutational step. The four mutations that improved function on LamB were thus potentiating mutations, in the sense that once they had been fixed, it became possible for the phage to access a mutation that allowed it to infect via OmpF. Thus potentiation is a phenomenon that arises as an epistatic consequence of the substitution of mutations in another context altogether. In the case of λ phage, all these mutations seem to have been beneficial. However, it seems plausible that in some cases potentiation could occur, because neutral (or even mildly deleterious) mutations that hitchhike to high frequency alongside beneficial mutations could be responsible for the potentiating effect.

Exaptation, as explained earlier, occurs when preexisting gene functions are co-opted for growth and reproduction under novel conditions. The important feature of exaptation is that it provides an initial set of genetic material on which further genetic modification can occur and, of more immediate importance, ensures that the population does not go extinct in the novel conditions. Exaptations exist for many metabolic functions because most enzymes are rather sloppy in their substrate specificity. Although they have one primary activity, they also possess a range of side activities that allow them to be co-opted into catalyzing other reactions if the appropriate enzyme is missing.

For bacteria, Bergthorsson, Andersson, and Roth (2007) have highlighted a number of studies showing that when a null mutation destroys activity of one gene, overexpression of non-cognate genes can often rescue function; the side activities of these genes are therefore supplying the missing function. For example, *Klebsiella aerogenes* and *E. coli* can be selected to use D-arabinose, a substrate that's rarely found in nature and that neither species uses at all. At the metabolic level, exaptation involved co-opting the enzyme L-fucose isomerase to convert D-arabinose to D-ribulose, a substrate that could then be fed into central metabolism by preexisting pathways. This and other examples of exaptation leading to novel metabolic functions in bacteria are reviewed by Mortlock (1983) and Hall (1989).

Further increases in fitness allowing a population to survive in the novel environment come about through various forms of gene product amplification. Mechanisms such as exon shuffling, gene translocation, mobile element transposition, and even chromosome rearrangements are all examples of how gene product amplification might happen. In microbial experiments, by far the most common means of increasing gene product is through loss-of-function mutations that deregulate gene expression and lead to constitutive production of an enzyme (Hall 1989) and gene duplications (Ohno 1970). Gene duplication in particular is known to occur in growing populations of bacteria at rates as high as 10^{-5} to 10^{-2} (Andersson and Hughes 2009), in part because genome replication rates are often higher than cell division rates during exponential growth. The end result is the same in both cases: make more of what you already have, even if what you have doesn't work so well. Examples of both mechanisms are easy to find in the literature; here are two. A cross-feeding polymorphism in *E. coli* grown on glucose was caused by a mutation in the *acs* operon that led to constitutive overexpression of acetyl Co-A synthetase and allowed the cross-feeding strain to efficiently scavenge acetate, a waste product of glucose metabolism (Kinnersley, Holben, and Rosenzweig 2009). Gene amplification was responsible for the recovery of fitness in *hemC* mutants of *Salmonella typhimurium* that confer resistance to the antimicrobial peptide protamine (Pränting and Andersson 2011). *hemC* makes an enzyme involved in the biosynthesis of heme (an iron-containing organic molecule that binds oxygen) and the resistant mutant contained a point mutation that compromised electron transport in this process. Interestingly, *hemC* itself was amplified, presumably resulting in more of the defective enzyme and allowing the genotype to grow faster.

Increases in population size lead to a reduced probability of extinction and a higher mutation supply rate. With more mutations available to selection, gene function can be modified and refined by selection, leading eventually to genetic and phenotypic divergence. Divergence unfolds through the evolution of fitness trade-offs, a subject discussed at length in Chapter 3.

Diversification ends here in asexual systems, with the evolution of niche specialization. However, in sexual systems it is reproductive isolation that marks the end point of diversification. Theories of ecological speciation see the genes that are substituted by selection when adapting to novel environments as those that also contribute to the evolution of reproductive isolation between populations. Evidence supporting this view comes from the yeast experiment mentioned earlier in this chapter that documented a Dobhzansky-Muller incompatibility resulting from the antagonistic, epistatic effects of mutations that evolved independently and so have never been tested together (Anderson et al. 2010; Dettman et al. 2007). The importance of Dobhzansky-Muller incompatibilities or other mechanisms for ecological speciation in generating reproductive isolation in higher organisms remains a subject of intense scrutiny (Nosil 2012). Interestingly, the limited genetic evidence to date suggests that reproductive isolation may sometimes be caused by genes that are independent of those causing adaptive divergence (Presgraves 2010). This topic deserves further attention in the experimental literature.

The PEAD(R) model comprises a general framework within which we can understand the sorts of genetic changes involved in diversification and how they connect to the ecology being experienced by an incipiently or actively diverging population. The evolution of striking instances of genetic novelty and innovation, as in the evolution of citrate use in *E. coli*, requires key innovations that can take a long time to evolve because they require a series of seemingly fortuitous genetic changes to occur that potentiate a lineage. Once a lineage becomes potentiated, exaptation becomes accessible to mutation, and diversification is expected to proceed as in the conventional EAD model. Rapid diversification like that seen in the *P. fluorescens* radiation, on the other hand, comes about because the founding genotype clearly has many of the key genetic elements already in place. The genetic machinery for producing a biofilm exists (exaptation) and is accessible in a single mutational step, meaning there is no need for potentiation. The mutations causing biofilm expression result from deregulation of key proteins (amplification) and have strongly antagonistic pleiotropic effects (diversification). The result is rapid adaptive radiation within just a few tens of generations.

The PEAD(R) model thus constitutes an expanded genetic theory of diversification that builds on the original EAD model to include potentiation and, for sexual lineages, reproductive isolation. Its main value is to help clarify the various genetic steps involved in diversification. The model also may be useful for interpreting why some diversification events appear to be so fast and others so slow. Rapid diversification occurs when a lineage is, if you like, pre-potentiated, in the sense that a novel function can be accessed by many single-step mutations. The *P. fluorescens* radiation is an example. On the other hand, striking examples of novelty and innovation probably take much longer and may often be rarer because, as in the evolution of citrate use in *E. coli,* they require potentiating mutations to occur in ways that make the evolution of novel function a highly contingent event.

CONCLUSIONS

Our understanding of the factors responsible for evolutionary diversification has so far been woefully one-sided. To say, for example, that consumer preferences are the source of divergent and possibly disruptive selection responsible for the wildly different microbial composition of the same fermented food products in a market in Zambia—even if we could verify it—tells only half of the story. It leaves out entirely the issue of the genetic causes underlying such diversification. Surely a general theory of diversification must be able to account for both ecology and genetics.

It's just now becoming possible to incorporate genetics into ecological theories of diversification. The combination of experimental evolution and whole genome sequencing stands to make important contributions to our understanding of the genetic causes of diversification and the genomic patterns resulting from divergent selection. Even so, we still have a long way to go. At the time of this writing, little in the way of published literature deals with the subject.

No doubt this situation will change in the next few years, as the costs of whole genome sequencing continue to decrease and we gain a better understanding of the genomics of adaptation to different environments. For the time being, some intriguing patterns have emerged from those studies that have been done:

1. Divergent selection reduces the probability of parallel evolution and generates genetic differentiation by fixing mutations with environment-specific effects. The rate of genetic divergence can be quantified: on average 1.5 more unique mutations get substituted every 100 generations due to divergent selection relative to the divergence of populations responding to similar selection pressures.

2. The more niche dimensions a lineage is required to adapt to, the more mutations become fixed. The implication is that multifarious selection may be more effective at generating genetic divergence than is more homogeneous selection.

3. Divergent selection leads to genomic regions of divergence that typically affect no more than about 10 percent of the genome.

4. Although based on just a single example, experiments with yeast have now confirmed that the same genes involved in adaptation to different environments can also cause reproductive isolation. This observation lends

direct experimental support to a key prediction of the theory of ecological speciation.

5. Potentiation is the evolution of a genetic background that makes it possible for a novel function to arise by a single mutation; it evolves through the fortuitous fixation of mutations, through either selection or hitchhiking.

6. Incorporating potentiation into the exaptation-amplification-divergence (EAD) model leads to a broader, more general model for the genetics of diversification. Each step in the model—potentiation, exaptation, amplification, divergence, and (for sexuals) reproductive isolation, PEAD(R) for short—defines a sequence of genetic steps involved in diversification.

7. The PEAD(R) model can account for variation among lineages in the rate of diversification as well as the extent to which diversification is associated with phenotypic innovation and novelty.

These results, based on what is admittedly very limited data, give us a first glimpse into what we might call the genomic natural history of diversification. On the whole, they fit well with our intuition. It's reassuring, for example, that divergent selection does in fact lead to the substitution of mutations specific to a given environment and that this seems an effective way of generating genetically distinct types. It's also not surprising that the number of mutations fixed should increase with the number of niche dimensions involved in selection. Whether such multifarious selection proves to be an effective means of generating rapid genetic differentiation and reproductive isolation remains to be seen.

Possibly more surprising is that the fraction of the genome showing evidence of divergent selection, in phage at least, is similar to that in incipiently diverging populations of higher organisms in nature: about 10 percent. We shouldn't make too much of this number just yet, but it does tell us that, like adaptation, divergence is most likely associated with just a few genes of fairly large effect and not the result of more modest divergence at many genes across the genome. Perhaps this result is an artifact of focusing on studies where selection was no doubt very strong and time scales quite short, being less than 1000 generations at the most. Still, the result remains a valuable first glimpse into the early stages of divergence.

When it comes to the identity and ecological characteristics of the genes involved in diversification, we know much less. This is simply because the relevant experiments and analysis have yet to be done. We know next to nothing, for example, about whether trade-offs in fitness typically evolve because of a single substitution with strongly antagonistically pleiotropic effects or whether they evolve through the substitution of multiple mutations. This sort of information can come only by combining the results of whole genome sequencing, which identifies the mutations that have evolved, with estimates of their fitness effects across environments. We need more experiments of this sort.

In any case, some general properties of the genetic changes involved in diversification do seem to emerge. These are captured in the PEAD(R) model, which is built around the core of another widely accepted model for the evolution of new gene functions—the EAD model. Whole genome sequencing has revealed that two other processes, one at either end of the EAD sequence, are involved.

At the front end, so to speak, is potentiation, the ability of a lineage to access the relevant genetic variation required to make use of preexisting gene functions that constitute the raw material on which natural selection works. We know little about how this happens, except that it involves adaptation to the native environment, although it possibly includes hitchhiking of nonadaptive mutations as well. These mutations fortuitously bring a lineage into a region of mutational space where it becomes possible to gain, through a single mutation, a novel function that allows it to invade a new environment. How and why this happens remains almost a complete mystery and represents a major challenge for the future.

At the back end is reproductive isolation. Adaptive divergence is commonly thought to be a necessary first step in generating reproductive isolation—a view that gains some credence from the one experiment, in yeast, that has examined it. However, the jury is still out on this topic because evidence from natural populations suggests that genes associated with hybrid dysfunction are not obviously related to adaptive divergence. For sexual systems at least, we may need an alternative theory of speciation to explain the final steps in lineage splitting.

A major advantage of seeing the genetic architecture of diversification through the lens of the PEAD(R) model is that it unites two features of diversification, the evolution of novelty and innovation on the one hand and niche specialization on the other, that are usually treated separately. PEAD(R) sees these as opposite ends of a sequence of steps that make a lineage more or less prone to diversify. A lineage that has ready access to single-step mutations that are strongly antagonistic is likely to diversify rapidly and in a way that produces no profoundly new phenotypes or functions. A lineage that must evolve an entirely novel gene function to gain access to a novel environment is unlikely to have ready access to the "right" mutations, and so it will take much longer to evolve. But when it does, that lineage will appear as a key innovation that spurs further diversification. One very productive avenue for future work will be to further evaluate the PEAD(R) model as a general explanation for evolutionary diversification.

CHAPTER 10

The Nature of Biodiversity

ᴄꙅ Much, if not all, biological research is directed toward answering one question: How has life become so diverse? Every discipline of biology has come to an explanation through the lens of its own interests and preoccupations. Habitat, area, time, behavior, sex, competition, coevolution, development, gene expression, climate, neutrality—the list goes on. Against such a background, it's no wonder efforts to construct a general theory of diversity often falter. In the end, the safest explanation always seems to be to point to Darwin and Wallace and say that natural selection is responsible. Surely we can do better.

This book attempts to do just that. The starting point is to recognize that evolutionary diversification has at its core two component processes, that of adaptation on the one hand and lineage splitting on the other. The text has tried to make the connection between these two processes more explicit. This task is best accomplished by studying adaptation and diversity directly, through experiment. This is why scientists seek answers in microbes, where it's possible to watch the evolutionary process unfold in real time and to track diversification in both phenotype and genotype along the way. The book provides an overview of the emerging field of microbial experimental evolution for what it tells us about the sorts of problems and questions related to adaptation and diversity that have preoccupied "real," muddy-boots biologists. This chapter sums up and assesses what microbial experimental evolution teaches us about the evolution of biodiversity.

A GENERAL THEORY OF DIVERSIFICATION

A general theory of diversification is one that accounts for both the ecology and the genetics of diversification. To date, evolutionary biology has emphasized the former and ignored the latter. In the past, this approach could be justified on both theoretical and practical grounds. The theoretical argument was that without an understanding of the broader principles of selection and how it generates and maintains diversity, any theory of diversification would be meaningless. Practically, the rationale has always been that identifying the genetic causes of adaptation and diversification has not, except in unusual circumstances, been possible.

Both justifications are now much harder to defend. Researchers in ecology and population genetics spent much of the twentieth century identifying the principles of selection and how it maintains diversity. More recently, we have supplemented this theory with a more complete account of adaptation that predicts not only the dynamics of substitution itself but also the magnitude of the fitness increases that result. Microbial experiments have played an essential role in testing this theory. Significant gaps remain, to be sure, particularly over how environmental variation affects the process of substitution, but these are likely to be overcome in the next few years. The result should be a fairly comprehensive and empirically grounded theory of adaptation. Together with an understanding of how effective selection can be at generating local adaptation, we will be in a strong position to provide an account of the evolutionary origins of adaptive diversification.

The more practical argument no longer holds water. The technical obstacles preventing the discovery of the genetic causes of adaptive change have largely been overcome. Every year, the cost of using next-generation sequencing techniques seems to fall. It's now possible to scan entire genomes for the genetic targets of selection. This approach is especially powerful when combined with laboratory selection experiments in microbes, because the genetic changes that arise over the course of a selection experiment can be comprehensively identified, catalogued, and investigated for their effects on adaptation and diversification. Thus it's now feasible, at least in microbial selection experiments, to integrate genetics directly into our ecological theories of diversification. So, what does such a theory look like?

The simplest version of the theory goes like this. On the ecological side, divergent selection, generated ultimately by ecological opportunity, or vacant niche space, results in adaptation that generates trade-offs in fitness across environments. Any factor that changes the extent of ecological opportunity or modulates the strength of divergent selection affects the rate and extent of diversification. Increases to both factors lead to rapid and spectacular diversification events like adaptive radiation. Decreases can slow or prevent diversification altogether. Microbial experiments provide no reason to question this interpretation, and in the process they have contributed immensely to the development of a more empirically robust theory of adaptation.

On the genetic side, we are now in a position to say with much more confidence that, yes, certain kinds of genes or genetic architectures are more likely to generate adaptation and even diversification. For adaptation, genes involved in regulating broad patterns of gene expression across the genome, especially in response to the stress associated with adaptation to novel conditions, are often observed to contribute to increases in fitness in microbial experiments. So too are the losses of genes (or gene functions) that are no longer needed in a novel environment—a process of domestication, as it were. These results suggest that the initial stages of adaptation to novel environments that are associated with diversification are likely to be fairly predictable, at least at the level of gene function. More refined predictions about the identity of specific genes, or even specific mutations, remain a challenge. For diversification, the key factor is the supply of mutations from which selection generates adaptive divergence. As a way of framing our thinking about how mutation supply depends on features of the genetic architecture, the preceding chapter introduced the potentiation, exaptation, amplification, divergence, and reproductive isolation [PEAD(R)] model. Any genetic change—a duplication, for example—that increases the availability of mutations while also increasing fitness in the novel environment will tend to predispose a lineage toward diversification.

Together, the ecological and genetic sides of the theory take us some way toward a more comprehensive view of how diversification occurs. They allow us, for example, to understand more clearly why some lineages have undergone rapid and extensive diversification such as adaptive radiations while others have not. Adaptive radiations are fairly unique situations, because they require an unusual alignment of factors on both the ecological and genetic sides. Ecologically, the conditions for adaptive radiation include abundant ecological opportunity and strong divergent or even disruptive selection, the latter contributed by ecological interactions like resource competition. On the genetic side, the ancestral lineage that gives rise to the radiation must be primed to do so. It cannot, for example, be so maladapted that it must go through a process of potentiation, waiting for multiple mutations to be substituted that, for whatever reason, permit it to access beneficial mutations relevant to divergence; this process takes too much time to be counted as an adaptive radiation. Rather, a lineage that undergoes adaptive radiation is more likely to be located toward the amplification-divergence end of the PEAD(R) spectrum, where an abundance of genetic variation allows niche specialization to evolve rapidly.

THE STRENGTH OF INFERENCE

It's reasonable to ask whether the microbial approach can tell us anything useful about adaptation and diversification in nature. For example, a serious limitation of using microbial systems to study adaptive radiation is that the model systems employed to date are all asexual, yet many of the radiations we care most about—the Galápagos finches, African Rift lake cichlids, Hawaiian silverswords—are sexual. Furthermore, the insights microbes have afforded into the genomics of adaptation, that most substitutions are adaptive, for example, can be understood only in the context of the parameter space in which the experiments are done: large, asexual populations subject to strong selection and usually (to date, at least) no migration. These issues, and no doubt many others, call into question the inferential strength of the systems being studied. There are, to this author's mind, four responses to that criticism.

The first is to recognize the limits to inference of the microbial enterprise. Microbial experiments, like other models, work best when they are used to evaluate the plausibility of a particular mechanism or to decide among competing mechanisms. Deciding which mechanism is actually driving adaptation and diversification in any natural system remains a task for field research. Collaborative efforts that integrate the insights gained from microbial microcosms into a broader research program will be necessary to make this happen. Those of us who do microbial experiments can help by always reporting key demographic parameters such as effective population sizes and mutation rates, so that it's absolutely clear which region of population genetics parameter space we are working in.

Second, although microbial systems studied to date are simple ones, this does not prevent more complex situations from being studied in the future. The effect of sexual reproduction on diversification is severely understudied, for example, yet it can in principle be examined directly by turning to facultatively sexual systems such as unicellular fungi and algae. Such systems would moreover enable investigators to ask questions about the final step in diversification, reproductive isolation. They also may prove useful for examining the role of sexual selection in adaptation to novel environments and diversification (Rogers and Greig 2009). Given the putative importance of sexual selection in many adaptive radiations, like the African cichlids, this is likely to be an especially powerful tool. A host of other factors have yet to be studied explicitly in the context of adaptive diversification. We know little, for example, about how ploidy levels, competition among multiple predators, or interspecific competition affect adaptation and diversification. Progress toward a general theory requires that these factors be addressed.

Third, it may be useful to start seeing the microbial approach as a means of not only testing hypotheses, but generating them too. The examples of potentiation in *Escherichia coli* stand out in this regard: this process would seem too far-fetched without solid experimental evidence to back it up, yet we now have at least two good experimental examples suggesting that it occurs. Another might be an observation recently made that, in a selection experiment in *Pseudomonas fluorescens* evolving in minimal medium with glucose as a sole energy source, two synonymous mutations evolved that have beneficial fitness effects (Bailey, Hinz, and Kassen 2014). The evidence we have is incontrovertible: placing these mutations in the genetic background of the ancestor, with no other genetic changes, increases fitness by about 6 and 9 percent, respectively. How frequent beneficial synonymous mutations are beyond the one example mentioned here remains unknown, although recent work in *Drosophila melanogaster* suggests that as many as 22 percent of synonymous sites are under strong purifying selection (Lawrie et al. 2013). If these results are at all typical of synonymous sites more generally, then we will be forced to question the value of some basic tools of molecular population genetics for estimating selection, like the ratio of non-synonymous to synonymous sites, and to reexamine our understanding of how these seemingly innocuous mutations affect fitness.

Fourth, there are many situations where adaptation and diversification of microbes are themselves intrinsically interesting. This book highlights a couple of these situations in its introductory vignettes—on chronic infections of the cystic fibrosis lung by *P. aeruginosa,* for example, or the microbial communities in Zambian fermented food products. The factors governing adaptation and diversification in these systems, and many others like them, are not well understood. The general principles laid out in this book can help provide a starting place to make sense of the evolutionary processes at work in these more natural systems. Real progress will not come, however, until microbiologists themselves start examining the dynamics of fitness and genetic variation in the systems they are most interested in and that are most relevant to human and environmental health.

SPECULATIONS AND DIRECTIONS

The insights gained from microbial experiments have taken us some way toward developing a more general theory of diversification. However, a number of compelling and important issues remain that microbial systems are uniquely positioned to help answer. Many of these topics were alluded to at various points throughout the book, but at least three issues stand out as major gaps in our knowledge.

The first is the long-term fate of diversity. Ecological and population genetics models for maintaining diversity typically do not allow for further adaptive evolution. Yet it's clear from a number of results in microbial systems—the so-called growth advantage in stationary phase (GASP) mutants that emerge in prolonged batch cultures of *E. coli* (Zambrano et al. 1993), the turnover of different niche specialist morphotypes in long-term batch cultures of *P. fluorescens* (Meyer et al. 2011), and the dynamics of adaptation and coexistence among niche specialist genotypes of *E. coli* in chemostats (Dykhuizen and Dean 2004)—that many populations continue to evolve even after diversity has arisen and persisted for some time. In some instances, continued evolution seems to lead to the loss of diversity, and in others diversity is maintained or even increases. What explains these different fates for diversity? Can diversity be maintained indefinitely, as the simplest models from population genetics and ecology assume? Or does continued adaptation change the nature of the ecological interactions that act to preserve diversity, leading to its eventual loss? It's even conceivable that nature becomes ever better at finding new niches, by partitioning old ones, for example, or through complicated and often unpredictable niche-creation mechanisms like cross-feeding. If so, then perhaps diversity inevitably increases, and the problem we seek to explain is not what maintains diversity, but why there isn't more of it. Answering these questions represents a major task for the experimental evolution enterprise in the next 10 years.

The relative importance of standing genetic variation as opposed to mutation in generating adaptive divergence also remains understudied. Most of the microbial experiments reviewed here focus exclusively on variation arising de novo through mutation. Although standing genetic variation is known to play an important role in driving adaptation as well, it's not clear how much it can constrain adaptive divergence, and for how long. Microbes provide a relatively easy way to get an answer. Genetically diverse founding populations can be easily constructed from crosses between strains of facultatively sexual species such as yeast. Investigators could then contrast the rate and extent of diversification in experimental lines that varied in the quantity of standing genetic variance.

Finally, the genomic details of adaptation and diversification deserve closer attention. Although we have come some way toward providing a fuller account of the connection between genotype and phenotype during adaptive diversification, there's a long way yet to go. How often is ecological divergence tied to reproductive isolation, for example, as it was in the yeast experiment (Anderson et al. 2010; Dettman et al. 2007)? Are the largest beneficial effects on fitness in one environment also those that contribute most to trade-offs in fitness in alternative environments? What role does epistasis play in governing the repeatability of adaptive evolution and diversification? Does potentiation, which governs the accessibility of mutations that allow diversification, help us understand why some lineages diversify more readily than others? What effects does recombination have on the identity and genomic location of genes substituted under selection? Does recombination more often tend to interfere with the evolution of niche specialization or promote it? Is mutation–selection balance the ultimate source of what we call "new" mutations during adaptation from an initially clonal population, or are these mutations genuinely de novo? Answering these questions requires finer-scale sampling of the genomic changes associated with adaptation and diversification, both in time and within a population. The technologies available to do this are now becoming available and are increasingly affordable. The challenge will be to use them in a way that allows us to do more than just genomic natural history; we need to be doing more in the way of hypothesis testing as well.

These three questions are some of the most exciting for twenty-first-century evolutionary biology. Answering them will bring us much closer to a comprehensive understanding of the ecological and genetic processes involved in adaptive diversification. A general theory of diversification that integrates ecology and genetics is most definitely within reach, and this text has provided an initial attempt at one. No doubt it will be revised and discarded in the future as more experiments are done.

These questions also strike close to the heart of this book, and the title of this chapter—the nature of biodiversity. What microbial experiments have shown us to date about biodiversity is that it almost inevitably evolves in some form in our experiments, and often in surprising and unexpected ways. Cross-feeding polymorphisms are a perfect example, as is the evolution of biofilm-forming genotypes that stick to the walls of glass microcosms and resist being washed away but end up coexisting with broth-colonizing types. Diversity seems to find a way to evolve in almost all systems, eventually, whether we notice it or not. Our challenge is to find the right combination of tools to identify diversity and make sense of it. This remains the paramount challenge for biodiversity research in the future, whether that research is done in the lab or in the field.

Microbial experiments have played an integral role in getting to this point by providing a glimpse into the ecology and genetics of a world that, though highly contrived, is one that we can begin, at least, to get a handle on. They are not the be-all and end-all of evolutionary biology by any means. Their worth ultimately will be validated if they help us make better predictions about how adaptation and diversification happen in the real world. Nevertheless, the opportunity they afford us to study the evolutionary process in real time, and the depth of insight they provide into this process, make them essential to the development of any general theory of diversity. For this reason, microbial experimental evolution is and will remain a valuable and important tool for evolutionary biologists in the future.

Literature Cited

Abrams, P. A. 2000. The Evolution of Predator-Prey Interactions: Theory and Evidence. *Annual Review of Ecology and Systematics* 31: 79–105.

Alto, B. W., and P. E. Turner. 2010. Consequences of Host Adaptation for Performance of Vesicular Stomatitis Virus in Novel Thermal Environments. *Evolutionary Ecology* 24: 299–315.

Anderson, J. B., J. Funt, D. A. Thompson, S. Prabhu, A. Socha, C. Sirjusingh, et al. 2010. Determinants of Divergent Adaptation and Dobzhansky-Muller Interaction in Experimental Yeast Populations. *Current Biology* 20: 1383–88.

Anderson, J. B., C. Sirjusingh, and N. Ricker. 2004. Haploidy, Diploidy and Evolution of Antifungal Drug Resistance in *Saccharomyces cerevisiae*. *Genetics* 168: 1915–23.

Andersson, D. I., and D. Hughes. 2009. Gene Amplification and Adaptive Evolution in Bacteria. *Annual Review of Genetics* 43: 167–95.

Andersson, D. I., and D. Hughes. 2010. Antibiotic Resistance and Its Cost: Is It Possible to Reverse Resistance? *Nature Reviews Microbiology* 8: 260–71.

Applebee, M. K., M. J. Herrgård, and B. Ø. Palsson. 2008. Impact of Individual Mutations on Increased Fitness in Adaptively Evolved Strains of *Escherichia coli*. *Journal of Bacteriology* 190: 5087–94.

Araya, C. L., C. Payen, M. J. Dunham, and S. Fields. 2010. Whole-Genome Sequencing of a Laboratory-Evolved Yeast Strain. *BMC Genomics* 11: 88.

Arbuthnott, D., and H. Rundle. 2013. Misalignment of Natural and Sexual Selection among Divergently Adapted *Drosophila melanogaster* populations. *Animal Behaviour* in press.

Armstrong, R. A., and R. McGehee. 1980. Competitive Exclusion. *American Naturalist* 115: 151–70.

Arnqvist, G., M. Edvardsson, U. Friberg, and T. Nilsson. 2000. Sexual Conflict Promotes Speciation in Insects. *Proceedings of the National Academy of Sciences USA* 97: 10460–64.

Atsumi, S., T.-Y. Wu, I. M. P. Machado, W.-C. Huang, P.-Y. Chen, M. Pellegrini, et al. 2010. Evolution, Genomic Analysis, and Reconstruction of Isobutanol Tolerance in *Eschierichia coli*. *Molecular Systems Biology* 6: 449. doi:10.1038/msb.2010.98.

Atwood, K. C., L. K. Schneider, and F. J. Ryan. 1951. Periodic Selection in *Escherichia coli*. *Proceedings of the National Academy of Sciences USA* 37: 146–55.

Bailey, S. F., J. R. Dettman, P. B. Rainey, and R. Kassen. 2013. Competition both Drives and Impedes Diversification in a Model Adaptive Radiation. *Proceedings of the Royal Society B: Biological Sciences* 280: 20131253.

Bailey, S. F., A. Hinz, and R. Kassen. 2014. Adaptive Synonymous Mutations in an Experimentally Evolved Population. *Nature Communications* (in press).

Bailey, S. F., and R. Kassen. 2012. Spatial Structure of Ecological Opportunity Drives Adaptation in a Bacterium. *American Naturalist* 180: 270–83.

Baker, M. D., P. M. Wolanin, and J. B. Stock. 2006. Signal Transduction in Bacterial Chemotaxis. *Bioessays* 28: 9–22.

Baltrus, D., K. Guillemin, and P. Philips. 2008. Natural Transformation Increases the Rate of Adaptation in the Human Pathogen *Helicobacter pylori*. *Evolution* 62: 39–49.

Bantinaki, E., R. Kassen, C. G. Knight, Z. Robinson, A. J. Spiers, and P. B. Rainey. 2007. Adaptive Divergence in Experimental Populations of *Pseudomonas fluorescens*. III. Mutational Origins of Wrinkly Spreader Diversity. *Genetics* 176: 441–53.

Barnosky, A. D. 2001. Distinguishing the Effects of the Red Queen and Court Jester on Miocene Mammal Evolution in the Northern Rocky Mountains. *Journal of Vertebrate Paleontology* 21: 172–85.

Barraclough, T. G., P. H. Harvey, and S. Nee. 1995. Sexual Selection and Taxonomic Diversity in Passerine Birds. *Proceedings of the Royal Society B: Biological Sciences* 259: 211–15.

Barrett, R. D. H., and G. Bell. 2006. The Dynamics of Diversification in Evolving *Pseudomonas* Populations. *Evolution* 60: 484–90.

Barrett, R. D. H., and H. E. Hoekstra. 2011. Molecular Spandrels: Tests of Adaptation at the Genetic Level. *Nature Reviews Genetics* 12: 767–80.

Barrett, R. D. H., R. C. MacLean, and G. Bell. 2005. Experimental Evolution of *Pseudomonas fluorescens* in Simple and Complex Environments. *American Naturalist* 166: 470–80.

Barrett, R. D. H., L. K. M'Gonigle, and S. P. Otto. 2006. The Distribution of Beneficial Mutant Effects Under Strong Selection. *Genetics* 174: 2071–79.

Barrick, J. E., and R. E. Lenski. 2009. Genome-wide Mutational Diversity in an Evolving Population of *Escherichia coli*. *Cold Spring Harbor Symposia on Quantitative Biology* 74: 119–29.

Barrick, J. E., D. S. Yu, S. H. Yoon, H. Jeong, T. K. Oh, D. Schneider, et al. 2009. Genome Evolution and Adaptation in a Long-Term Experiment with *Escherichia coli*. *Nature* 461: 1243–47.

Bataillon, T., T. Zhang, and R. Kassen. 2011. Cost of Adaptation and Fitness Effects of Beneficial Mutations in *Pseudomonas fluorescens*. *Genetics* 189: 939–49.

Beaumont, H. J. E., J. Gallie, C. Kost, G. C. Ferguson, and P. B. Rainey. 2009. Experimental Evolution of Bet Hedging. *Nature* 462: 90–93.

Beisel, C. J., D. R. Rokyta, H. A. Wichman, and P. Joyce. 2007. Testing the Extreme Value Domain of Attraction for Distributions of Beneficial Fitness Effects. *Genetics* 176: 2441–49.

Bell, G. 1982. *The Masterpiece of Nature: the Evolution and Genetics of Sexuality*. London: Croom Helm.

———.1997a. The Evolution of the Life Cycle of Brown Seaweeds. *Biological Journal of the Linnean Society* 60: 21–38.

———. 1997b. Experimental Evolution in *Chlamydomonas*. I. Short-Term Selection in Uniform and Diverse Environments. *Heredity* 78: 490–97.

———. 2005. Experimental Sexual Selection in *Chlamydomonas*. *Journal of Evolutionary Biology* 18: 722–34.

———. 2008. *Selection: The Mechanism of Evolution* (2nd ed.). Oxford: Oxford University Press.

Bell, G., and A. Gonzalez. 2009. Evolutionary Rescue Can Prevent Extinction Following Environmental Change. *Ecology Letters* 12: 942–48.

———. 2011. Adaptation and Evolutionary Rescue in Metapopulations Experiencing Environmental Deterioration. *Science* 332: 1327–30.

Bell, G., and X. Reboud. 1997. Experimental Evolution in *Chlamydomonas*. II. Genetic Variation in Strongly Contrasted Environments. *Heredity* 78: 498–506.

Belotte, D., J.-B. Curien, R. C. MacLean, and G. Bell. 2003. An Experimental Test of Local Adaptation in Soil Bacteria. *Evolution* 57: 27–36.

Benmayor, R., A. Buckling, M. B. Bonsall, M. A. Brockhurst, and D. J. Hodgson. 2008. The Interactive Effects of Parasites, Disturbance, and Productivity on Experimental Adaptive Radiations. *Evolution* 62: 467–77.

Bennett, A. F., and R. E. Lenski. 1993. Evolutionary Adaptation to Temperature. II. Thermal Niches of Experimental Lines of *Escherichia coli*. *Evolution* 47: 1–12.

———. 2007. An Experimental Test of Evolutionary Trade-offs during Temperature Adaptation. *Proceedings of the National Academy of Sciences USA* 104 Suppl: 8649–54.

Bennett, A. F., R. E. Lenski, and J. E. Mittler. 1992. Evolutionary Adaptation to Temperature. I. Fitness Responses of *Escherichia coli* to Changes in Its Thermal Environment. *Evolution* 46: 16–30.

Benton, M. J. 1995. Diversification and Extinction in the History of Life. *Science* 268: 52–58.

———. 2009. The Red Queen and the Court Jester: Species Diversity and the Role of Biotic and Abiotic Factors through Time. *Science* 323: 728–32.

———. 2010. New Take on the Red Queen. *Nature* 463: 306–7.

Benton, M. J., and B. C. Emerson. 2007. How Did Life Become So Diverse? The Dynamics of Diversification According to the Fossil Record and Molecular Phylogenetics. *Palaeontology* 50: 23–40.

Bérénos, C., K. M. Wegner, and P. Schmid-Hempel. 2011. Antagonistic Coevolution with Parasites Maintains Host Genetic Diversity: An Experimental Test. *Proceedings. of the Royal Society B: Biological Sciences* 278: 218–24.

Bergthorsson, U., D. I. Andersson, and J. R. Roth. 2007. Ohno's Dilemma: Evolution of New Genes under Continuous Selection. *Proceedings of the National Academy of Sciences USA* 104: 17004–9.

Betancourt, A. J. 2009. Genomewide Patterns of Substitution in Adaptively Evolving Populations of the RNA Bacteriophage MS2. *Genetics* 181: 1535–44.

Blount, Z. D., J. E. Barrick, C. J. Davidson, and R. E. Lenski. 2012. Genomic Analysis of a Key Innovation in an Experimental *Escherichia coli* Population. *Nature* 489: 513–18.

Blount, Z. D., C. Z. Borland, and R. E. Lenski. 2008. Historical Contingency and the Evolution of a Key Innovation in an Experimental Population of *Escherichia coli*. *Proceedings of the National Academy of Sciences USA* 105: 7899–906.

Blows, M. W. 2007. A Tale of Two Matrices: Multivariate Approaches in Evolutionary Biology. *Journal of Evolutionary Biology* 20: 1–8.

Blows, M. W., and A. A. Hoffmann. 2005. A Reassessment of Genetic Limits to Evolutionary Change. *Ecology* 86: 1371–84.

Bohannan, B. J. M., and R. E. Lenski. 2000. Linking Genetic Change to Community Evolution: Insights from Studies of Bacteria and Bacteriophage. *Ecology Letters* 3: 362–77.

Bohren, B., W. G. Hill, and A. Robertson. 1966. Some Observations on Asymmetrical Correlated Responses to Selection. *Genetical Research* 7: 44–57.

Bolnick, D. I. 2001. Intraspecific Competition Favors Niche Width Expansion in *Drosophila melanogaster*. *Nature* 410: 463–66.

Bolnick, D., T. Ingram, W. Stutz, L. Snowberg, O. Lau, and J. Paull. 2010. Ecological Release from Interspecific Competition Leads to Decoupled Changes in Population and Individual Niche Width. *Proceedings of the Royal Society B: Biological Sciences* 277: 1789–97.

Boughman, J. W. 2001. Divergent sexual selection enhances reproductive isolation in sticklebacks. *Nature* 411: 944–48.

Brockhurst, M. A., A. Buckling, and P. B. Rainey. 2005. The Effect of a Bacteriophage on Diversification of the Opportunistic Bacterial Pathogen, *Pseudomonas aeruginosa*. *Proceedings of the Royal Society B: Biological Sciences* 272: 1385–91.

Brockhurst, M. A., N. Colegrave, D. J. Hodgson, and A. Buckling. 2007a. Niche Occupation Limits Adaptive Radiation in Experimental Microcosms. *PloS One* 2: e193.

Brockhurst, M. A., A. D. Morgan, A. Fenton, and A. Buckling. 2007b. Experimental Coevolution with Bacteria and Phage: The *Pseudomonas Fluorescens*–Phi2 Model System. *Infection, Genetics and Evolution* 7: 547–52.

Brockhurst, M. A., N. Colegrave, and D. E. Rozen. 2011. Next-Generation Sequencing as a Tool to Study Microbial Evolution. *Molecular Ecology* 20: 972–80.

Brockhurst, M. A., M. E. Hochberg, T. Bell, and A. Buckling. 2006. Character Displacement Promotes Cooperation in Bacterial Biofilms. *Current Biology* 16: 2030–34.

Brockhurst, M. A., P. B. Rainey, and A. Buckling. 2004. The Effect of Spatial Heterogeneity and Parasites on the Evolution of Host Diversity. *Proceedings of the Royal Society B: Biological Sciences* 271: 107–11.

Brown, C. J., K. M. Todd, and R. F. Rosenzweig. 1998. Multiple Duplications of Yeast Hexose Transport Genes in Response to Selection in a Glucose-Limited Environment. *Molecular Biology and Evolution* 15: 931–42.

Brusatte, S. L., S. J. Nesbitt, R. B. Irmis, R. J. Butler, M. J. Benton, and M. A. Norell. 2010. The Origin and Early Radiation of Dinosaurs. *Earth-Science Reviews* 101: 68–100.

Buckling, A., R. Kassen, G. Bell, and P. B. Rainey. 2000. Disturbance and Diversity in Experimental Microcosms. *Nature* 408: 961–64.

Buckling, A., and P. B. Rainey. 2002a. Antagonistic Coevolution between a Bacterium and a Bacteriophage. *Proceedings of the Royal Society B: Biological Sciences* 269: 931–36.

———. 2002b. The Role of Parasites in Sympatric and Allopatric Host Diversification. *Nature* 420: 496–99.

Buckling, A., M. A. Wills, and N. Colegrave. 2003. Adaptation Limits Diversification of Experimental Bacterial Populations. *Science* 302: 2107–9.

Bull, J. J., M. R. Badgett, H. A. Wichman, J. P. Huelsenbeck, D. M. Hillis, A. Gulati, et al. 1997. Exceptional Convergent Evolution in a Virus. *Genetics* 147: 1497–507.

Bull, J. J., M. R. Badgett, and H. A. Wichman. 2000. Big-Benefit Mutations in a Bacteriophage Inhibited with Heat. *Molecular Biology and Evolution* 17:942–50.

Bull, J. J., M. R. Badgett, R. Springman, and I. J. Molineux. 2004. Genome Properties and the Limits of Adaptation in Bacteriophages. *Evolution* 58: 692–701.

Bull, J. J., R. H. Heineman, and C. O. Wilke. 2011. The Phenotype-Fitness Map in Experimental Evolution of Phages. *PLoS ONE* 6: e27796.

Burch, C. L., and L. Chao. 1999. Evolution by Small Steps and Rugged Landscapes in the RNA Virus Phi6. *Genetics* 151: 921–27.

Byrne, K., and R. A. Nichols. 1999. *Culex pipiens* in London Underground Tunnels: Differentiation between Surface and Subterranean Populations. *Heredity* 82: 7–15.

Campos, P. R., P. S. C. Neto, V. M. de Oliveira, and I. Gordo. 2008. Environmental Heterogeneity Enhances Clonal Interference. *Evolution* 62: 1390–99.

Candolin, U., and J. Heuschele. 2008. Is Sexual Selection Beneficial during Adaptation to Environmental Change? *Trends in Ecology and Evolution* 23: 446–52.

Carroll, S. B., J. K. Grenier, and S. D. Weatherbee. 2001. *From DNA to Diversity: Molecular Genetics and the Evolution of Animal Design.* Malden, MA: Blackwell Scientific.

Castle, W. E. 1921. An Improved Method of Estimating the Number of Genetic Factors Concerned in Cases of Blending Inheritance. *Science* 54: 223.

Chao, L., B. R. Levin, and F. M. Stewart. 1977. A Complex Community in a Simple Habitat: An Experimental Study with Bacteria and Phage. *Ecology* 58: 369–78.

Charusanti, P., T. M. Conrad, E. M. Knight, K. Venkataraman, N. L. Fong, B. Xie, et al. 2010. Genetic Basis of Growth Adaptation of *Eschirichia coli* after Deletion of *pgi*, a Major Metabolic Gene. *PLoS Genetics* 6: e1001186.

Chase, J. M., and M. A. Leibold. 2003. *Ecological Niches: Linking Classical and Contemporary Approaches.* Chicago: University of Chicago Press.

Chesson, P. 2000. Mechanisms of Maintenance of Species Diversity. *Annual Review of Ecology and Systematics* 31: 343–66.

Chevin, L.-M., G. Martin, and T. Lenormand. 2010. Fisher's Model and the Genomics of Adaptation: Restricted Pleiotropy, Heterogenous Mutation, and Parallel Evolution. *Evolution* 64: 3213–31.

Chou, H.-H., H.-C. Chiu, N. F. Delaney, D. Segrè, and C. J. Marx. 2011. Diminishing Returns Epistasis among Beneficial Mutations Decelerates Adaptation. *Science* 332: 1190–92.

Chow, Y. K., M. S. Hirsch, D. P. Merrill, L. J. Bechtel, J. J. Eron, J. C. Kaplan, et al. 1993. Use of Evolutionary Limitations of HIV-1 Multidrug Resistance to Optimize Therapy. *Nature* 361: 650–54.

Ciota, A. T., A. O. Lovelace, K. A. Ngo, A. N. Le, J. G. Maffei, M. A. Franke, et al. 2007. Cell-Specific Adaptation of Two Flaviviruses Following Serial Passage in Mosquito Cell Culture. *Virology* 357: 165–74.

Coelho, S. M., A. F. Peters, B. Charrier, D. Roze, C. Destombe, M. Valero, et al. 2007. Complex Life Cycles of Multicellular Eukaryotes: New Approaches Based on the Use of Model Organisms. *Gene* 406: 152–70.

Coffey, L. L., N. Vasilakis, A. C. Brault, A. M. Powers, F. Tripet, and S. C. Weaver. 2008. Arbovirus Evolution in vivo Is Constrained by Host Alternation. *Proceedings of the National Academy of Sciences USA* 105: 6970–75.

Colegrave, N. 2002. Sex Releases the Speed Limit on Evolution. *Nature* 420: 664–66.

Colegrave, N., O. Kaltz, and G. Bell. 2002. The Ecology and Genetics of Fitness in *Chlamydomonas*. VIII. The Dynamics of Adaptation to Novel Environments After a Single Episode of Sex. *Evolution* 56: 14-21.

Collins, S. 2011. Competition Limits Adaptation and Productivity in a Photosynthetic Alga at Elevated CO_2. *Proceedings of the Royal Society B: Biological Sciences* 278: 247–55.

Collins, S., and G. Bell. 2004. Phenotypic Consequences of 1,000 Generations of Selection at Elevated CO_2 in a Green Alga. *Nature* 431: 566–69.

Collins, S., and J. de Meaux. 2009. Adaptation to Different Rates of Environmental Change in *Chlamydomonas*. *Evolution* 63: 2952–65.

Collins, S., J. de Meaux, and C. Acquisti. 2007. Adaptive Walks Toward a Moving Optimum. *Genetics* 176: 1089–99.

Conant, G. C., and K. H. Wolfe. 2008. Turning a Hobby into a Job: How Duplicated Genes Find New Functions. *Nature Reviews Genetics* 9: 938–50.

Conrad, T. M., M. Frazier, A. R. Joyce, B.-K. Cho, E. M. Knight, N. E. Lewis, et al. 2010. RNA Polymerase Mutants Found through Adaptive Evolution Reprogram *Escherichia coli* for Optimal Growth in Minimal Media. *Proceedings of the National Academy of Sciences USA* 107: 20500–20505.

Conrad, T. M., A. R. Joyce, M. K. Applebee, C. L. Barrett, B. Xie, Y. Gao, et al. 2009. Whole-Genome Resequencing of *Escherichia coli* K-12 MG1655 Undergoing Short-Term Laboratory Evolution in Lactate Minimal Media Reveals Flexible Selection of Adaptive Mutations. *Genome Biology* 10: R118.

Cooper, L. A., and T. W. Scott. 2001. Differential Evolution of Eastern Equine Encephalitis Virus Populations in Response to Host Cell Type. *Genetics* 157: 1403–12.

Cooper, T. F. 2007. Recombination Speeds Adaptation by Reducing Competition between Beneficial Mutations in Populations of *Escherichia coli*. *PLoS Biology* 5: e225.

Cooper, T. F, and R. E. Lenski. 2010. Experimental Evolution with *E. coli* in Diverse Resource Environments. I. Fluctuating Environments Promote Divergence of Replicate Populations. *BMC Evolutionary Biology* 10: 11.

Cooper, T. F., D. E. Rozen, and R. E. Lenski. 2003. Parallel Changes in Gene Expression after 20,000 Generations of Evolution in *Escherichia coli*. *Proceedings of the National Academy of Sciences* 100: 1072–77.

Cooper, V. S. 2002. Long-Term Experimental Evolution in *Escherichia coli*. X. Quantifying the Fundamental and Realized Niche. *BMC Evolutionary Biology* 2: 12.

Cooper, V. S., A. F. Bennett, and R. E. Lenski. 2001. Evolution of Thermal Dependence of Growth Rate of *Escherichia coli* Populations during 20,000 Generations in a Constant Environment. *Evolution* 55: 889–96.

Cooper, V. S., and R. E. Lenski. 2000. The Population Genetics of Ecological Specialization in Evolving *Escherichia coli* Populations. *Nature* 407: 736–39.

Coyne, J. A., and H. A. Orr. 2004. *Speciation*. Sunderland, MA: Sinauer Associates.

Crisp, M. D., M. T. K. Arroyo, L. G. Cook, M. A. Gandolfo, G. J. Jordan, M. S. McGlone, et al. 2009. Phylogenetic Biome Conservatism on a Global Scale. *Nature* 458: 754–56.

Crozat, E., N. Philippe, R. E. Lenski, J. Geiselmann, and D. Schneider. 2005. Long-Term Experimental Evolution in *Escherichia coli*. XII. DNA Topology as a Key Target of Selection. *Genetics* 169: 523–32.

Crozat, E., C. Winkworth, J. Gaffé, P. F. Hallin, M. A. Riley, R. E. Lenski, et al. 2010. Parallel Genetic and Phenotypic Evolution of DNA Superhelicity in Experimental Populations of *Escherichia coli*. *Molecular Biology and Evolution* 27: 2113–28.

Cuevas, J. M., P. Domingo-Calap, and R. Sanjuán. 2012. The Fitness Effects of Synonymous Mutations in DNA and RNA Viruses. *Molecular Biology and Evolution* 29: 17–20.

Cuevas, J. M., A. Moya, and S. F. Elena. 2003. Evolution of RNA Virus in Spatially Structured Heterogeneous Environments. *Journal of Evolutionary Biology* 16: 456–66.

Dallinger, W. H. 1887. The President's Address. *Journal of the Royal Microscopical Society* 7: 184–99.

Da Silva, J., and G. Bell. 1992. The Ecology and Genetics of Fitness in *Chlamydomonas*. VI. Antagonism between Natural Selection and Sexual Selection. *Proceedings of the Royal Society B: Biological Sciences* 249: 227–33.

Day, T. 2000. Competition and the Effect of Spatial Resource Heterogeneity on Evolutionary Diversification. *American Naturalist* 155: 790–803.

Day, T., and K. A. Young. 2004. Competitive and Facilitative Evolutionary Diversification. *BioScience* 54: 101–9.

De Paepe, M., V. Gaboriau-Routhiau, D. Rainteau, S. Rakotobe, F. Taddei, and N. Cerf-Bensussan. 2011. Trade-off between Bile Resistance and Nutritional Competence Drives *Escherichia coli* Diversification in the Mouse Gut. *PLoS Genetics* 7: e1002107.

De Varigny, H. 1892. *Experimental Evolution*. London: Macmillan.

Débarre, F., O. Ronce, and S. Gandon. 2013. Quantifying the Effects of Migration and Mutation on Adaptation and Demography in Spatially Heterogeneous Environments. *Journal of Evolutionary Biology* 26: 1185–202.

Dempster, E. R. 1955. Maintenance of Genetic Heterogeneity. *Cold Spring Harbor Symposia on Quantitative Biology* 20: 25–32.

Desai, M. M., and D. S. Fisher. 2007. Beneficial Mutation Selection Balance and the Effect of Linkage on Positive Selection. *Genetics* 176: 1759–98.

Dettman, J. R., J. B. Anderson, and L. M. Kohn. 2008. Divergent Adaptation Promotes Reproductive Isolation among Experimental Populations of the Filamentous Fungus *Neurospora*. *BMC Evolutionary Biology* 8: 35.

Dettman, J. R., N. Rodrigue, A. Melnyk, A. Wong, S. Bailey, and R. Kassen. 2012. Evolutionary Insight from Whole-Genome Sequencing of Experimentally Evolved Microbes. *Molecular Ecology* 21: 2058–77.

Dettman, J. R., C. Sirjusingh, L. M. Kohn, and J. B. Anderson. 2007. Incipient Speciation by Divergent Adaptation and Antagonistic Epistasis in Yeast. *Nature* 447: 585–88.

De Visser, J. A. G. M., C. W. Zeyl, P. J. Gerrish, J. L. Blanchard, and R. E. Lenski. 1999. Diminishing Returns from Mutation Supply Rate in Asexual Populations. *Science* 283: 404–6.

Dickerson, G. E. 1955. Genetic Slippage in Response to Selection for Multiple Objectives. *Cold Spring Harbor Symposia on Quantitative Biology* 20: 213–24.

Doebeli, M., and U. Dieckmann. 2000. Evolutionary Branching and Sympatric Speciation Caused by Different Types of Ecological Interactions. *American Naturalist* 156: S77–S101.

Drake, J. W. 1991. A Constant Rate of Spontaneous Mutation in DNA-Based Microbes. *Proceedings of the National Academy of Sciences USA* 88: 7160–64.

Duffy, M., C. E. Brassil, S. R. Hall, A. J. Tessier, C. E. Cáceres, and J. K. Conner. 2008. Parasite-Mediated Disruptive Selection in a Natural *Daphnia* Population. *BMC Evolutionary Biology* 8: 80.

Duffy, S., P. E. Turner, and C. L. Burch. 2006. Pleiotropic Costs of Niche Expansion in the RNA Bacteriophage Phi 6. *Genetics* 172: 751–57.

Dujon, B. 2010. Yeast Evolutionary Genomics. *Nature Reviews Genetics* 11: 512–24.

Dunham, M. J., H. Badrane, T. Ferea, J. Adams, P. O. Brown, F. Rosenzweig, et al. 2002. Characteristic Genome Rearrangements in Experimental Evolution of *Saccharomyces cerevisiae*. *Proceedings of the National Academy of Sciences USA* 99: 16144–49.

Dykhuizen, D. E. 1990. Experimental Studies of Natural Selection in Bacteria. *Annual Review of Ecology and Systematics* 21: 373–98.

Dykhuizen, D. E., and A. M. Dean. 2004. Evolution of Specialists in an Experimental Microcosm. *Genetics* 167: 2015–26.

Ehrlich, P. R., and P. H. Raven. 1964. Butterflies and Plants: A Study in Coevolution. *Evolution* 18: 586–608.

Elena, S. F., and R. E. Lenski. 1997. Long-Term Experimental Evolution in *Escherichia coli*. VII. Mechanisms Maintaining Genetic Variability within Populations. *Evolution* 51: 1058–67.

Emmerson, A. M., and A. M. Jones. 2003. The Quinolones: Decades of Development and Use. *Journal of Antimicrobial Chemotherapy* 51 Suppl. 1: 13–20.

Emerson, B. C., and N. Kolm. 2005. Species Diversity Can Drive Speciation. *Nature* 434: 1015–17.

ENCODE Project Consortium. 2012. An Integrated Encyclopedia of DNA Elements in the Human Genome. *Nature* 489: 57–74.

Engelmann, K. E., and C. D. Schlichting. 2005. Coarse- versus Fine-Grained Water Stress in *Arabidopsis thaliana* (Brassicaceae). *American Journal of Botany* 92: 101–6.

Erwin, D. H. 2007. Disparity: Morphological Pattern and Developmental Context. *Paleontology* 50: 57–73.

———. 2008. Macroevolution of Ecosystem Engineering, Niche Construction and Diversity. *Trends in Ecology & Evolution* 23: 304–10.

———. 2009. Climate as a Driver of Evolutionary Change. *Current Biology* 19: R575–83.

Eyre-Walker, A., and P. D. Keightley. 2007. The Distribution of Fitness Effects of New Mutations. *Nature Reviews Genetics* 8: 610–18.

Ezard, T. H. G., T. Aze, P. N. Pearson, and A. Purvis. 2011. Interplay between Changing Climate and Species' Ecology Drives Macroevolutionary Dynamics. *Science* 332: 349–51.

Falconer, D. S. 1981. *Introduction to Quantitative Genetics* (2nd ed.). London: Longman.

———. 1990. Selection in Different Environments: Effects on Environmental Sensitivity (Reaction Norm) and on Mean Performance. *Genetical Research* 56: 57–70.

Federenko, A. Y., and B. G. Shepherd. 1986. Review of Salmon Transplant Procedures and Suggested Transplant Guidelines. *Canadian Technical Report of Fisheries and Aquatic Sciences* 1479.

Felsenstein, J. 1976. The Theoretical Population Genetics of Variable Selection and Migration. *Annual Review of Genetics* 10: 253–80.

Ferenci, T. 2005. Maintaining a Healthy SPANC Balance through Regulatory and Mutational Adaptation. *Molecular Microbiology* 57: 1–8.

Ferenci, T., and B. Spira. 2007. Variation in Stress Responses within a Bacterial Species and the Indirect Costs of Stress Resistance. *Annals of the New York Academy of Sciences* 1113: 105–13.

Finkel, S. E., and R. Kolter. 1999. Evolution of Microbial Diversity during Prolonged Starvation. *Proceedings of the National Academy of Sciences USA* 96: 4023–27.

Finlay, K. W., and G. N. Wilkinson. 1963. The Analysis of Adaptation in a Plant-Breeding Programme. *Australian Journal of Agricultural Research* 14: 742–54.

Fisher, R. A. 1930. *The Genetical Theory of Natural Selection.* Oxford: Clarendon.

Folkesson, A., L. Jelsbak, L. Yang, H. K. Johansen, O. Ciofu, N. Høiby, et al. 2012. Adaptation of *Pseudomonas aeruginosa* to the Cystic Fibrosis Airway: An Evolutionary Perspective. *Nature Reviews Microbiology* 10: 841–51.

Fong, S. S., A. R. Joyce, and B. Ø. Palsson. 2005. Parallel Adaptive Evolution Cultures of *Escherichia coli* Lead to Convergent Growth Phenotypes with Different Gene Expression States. *Genome Research* 15: 1365–72.

Fonseca, D. M., N. Keyghobadi, C. A. Malcolm, C. Mehmet, F. Schaffner, M. Mogi, et al. 2004. Emerging Vectors in the *Culex pipiens* Complex. *Science* 303: 1535–38.

Foote, M. 1993. Discordance and Concordance Between Morphological and Taxonomic Diversity. *Paleobiology* 19: 185–204.

———. 2007. Symmetric Waxing and Waning of Marine Invertebrate Genera. *Paleobiology* 33: 517–29.

Francino, M. P. 2005. An Adaptive Radiation Model for the Origin of New Gene Functions. *Nature Genetics* 37: 573–77.

Fraser, D. J., L. K. Weir, L. Bernatchez, M. M. Hansen, and E. B. Taylor. 2011. Extent and Scale of Local Adaptation in Salmonid Fishes: Review and Meta-analysis. *Heredity* 106: 404–20.

Friesen, M. L., G. Saxer, M. Travisano, and M. Doebeli. 2004. Experimental Evidence for Sympatric Ecological Diversification Due to Frequency-Dependent Competition in *Escherichia coli. Evolution* 58: 245–60.

Friman, V.-P., T. Hiltunen, J. Laakso, and V. Kaitala. 2008. Availability of Prey Resources Drives Evolution of Predator-Prey Interaction. *Proceedings of the Royal Society B: Biological Sciences* 275: 1625–33.

Fukami, T., H. J. E. Beaumont, X.-X. Zhang, and P. B. Rainey. 2007. Immigration History Controls Diversification in Experimental Adaptive Radiation. *Nature* 446: 436–39.

Futuyma, D. J., and G. Moreno. 1988. The Evolution of Ecological Specialization. *Annual Review of Ecology and Systematics* 19: 207–33.

Gallet, R., S. Alizon, P.-A. Comte, A. Gutierrez, F. Depaulis, M. van Baalen, et al. 2007. Predation and Disturbance Interact to Shape Prey Species Diversity. *American Naturalist* 170: 143–54.

Gause, G. F. 1934. *The Struggle for Existence.* Baltimore: Williams & Wilkins.

Gavrilets, S., and A. Vose. 2005. Dynamic Patterns of Adaptive Radiation. *Proceedings of the National Academy of Sciences USA* 102: 18040–45.

Gerrish, P. J., and R. E. Lenski. 1998. The Fate of Competing Beneficial Mutations in an Asexual Population. *Genetica* 102/103: 127–44.

Gerstein, A. C. 2013. Mutational Effects Depend on Ploidy Level: All Else Is Not Equal. *Biology Letters* 9: 20120614.

Gerstein, A. C., H.-J. E. Chun, A. Grant, and S. P. Otto. 2006. Genomic Convergence Toward Diploidy in *Saccharomyces cerevisiae. PLoS Genetics* 2: e145.

Gerstein, A. C., L. Cleathero, M. Mandegar, and S. P. Otto. 2011. Haploids Adapt Faster than Diploids Across a Range of Environments. *Journal of Evolutionary Biology* 24: 531–40.

Gerstein, A. C., D. S. Lo, and S. P. Otto. 2012. Parallel Genetic Changes and Nonparallel Gene-Environment Interactions Characterize the Evolution of Drug Resistance in Yeast. *Genetics* 192: 241–52.

Gerstein, A. C., and S. P. Otto. 2009. Ploidy and the Causes of Genomic Evolution. *Journal of Heredity* 100: 571–81.

Gifford, D. R., and S. E. Schoustra. 2013. Modelling Colony Population Growth in the Filamentous Fungus *Aspergillus nidluans*. *Journal of Theoretical Biology* 320: 124–30.

Gifford, D. R., S. E. Schoustra, and R. Kassen. 2011. The Length of Adaptive Walks Is Insensitive to Starting Fitness in *Aspergillus nidulans*. *Evolution* 65: 3070–78.

Gillespie, J. H. 1984. Molecular Evolution over the Mutational Landscape. *Evolution* 38: 1116–29.

Gillespie, R. 2004. Community Assembly through Adaptive Radiation in Hawaiian Spiders. *Science* 303: 356–59.

Giraud, A., S. Arous, M. De Paepe, V. Gaboriau-Routhiau, J.-C. Bambou, S. Rakotobe, et al. 2008. Dissecting the Genetic Components of Adaptation of *Escherichia coli* to the Mouse Gut. *PLoS Genetics* 4: e2.

Goddard, M. R., H. C. Godfray, and A. Burt. 2005. Sex Increases the Efficacy of Natural Selection in Experimental Yeast Populations. *Nature* 434: 636–40.

Görke, B., and J. Stülke. 2008. Carbon Catabolite Repression in Bacteria: Many Ways to Make the Most Out of Nutrients. *Nature Reviews Microbiology* 6: 613–24.

Gould, S. J. 1989. *Wonderful Life.* New York: W. W. Norton.

Grant, B. R., and P. R. Grant. 2003. What Darwin's Finches Can Teach Us About the Evolutionary Origin and Regulation of Biodiversity. *BioScience* 53: 965–75.

Greischar, M. A., and B. Koskella. 2007. A Synthesis of Experimental Work on Parasite Local Adaptation. *Ecology Letters* 10: 418–34.

Gresham, D., B. Curry, A. Ward, D. B. Gordon, L. Brizuela, L. Kruglyak, et al. 2010. Optimized Detection of Sequence Variation in Heterozygous Genomes Using DNA Microarrays with Isothermal-Melting Probes. *Proceedings of the National Academy of Sciences USA* 107: 1482–87.

Gresham, D., M. M. Desai, C. M. Tucker, H. T. Jenq, D. A. Pai, A. Ward, et al. 2008. The Repertoire and Dynamics of Evolutionary Adaptations to Controlled Nutrient-Limited Environments in Yeast. *PLoS Genetics* 4: e1000303.

Gresham, D., D. M. Ruderfer, S. C. Pratt, J. Schacherer, M. J. Dunham, D. Botstein, et al. 2006. Genome-wide Detection of Polymorphisms at Nucleotide Resolution with a Single DNA Microarray. *Science* 311: 1932–36.

Habets, M. G. J. L., D. E. Rozen, R. F. Hoekstra, and J. A. G. M. de Visser. 2006. The Effect of Population Structure on the Adaptive Radiation of Microbial Populations Evolving in Spatially Structured Environments. *Ecology Letters* 9: 1041–48.

Hadany, L., and S. P. Otto. 2007. The Evolution of Condition-Dependent Sex in the Face of High Costs. *Genetics* 176: 1713–27.

Hairston Jr., N. G., S. Ellner, and C. M. Kearns. 1996. Overlapping Generations: The Storage Effect and the Maintenance of Biotic Diversity. In *Population Dynamics in Ecological Space and Time,* ed. O. Rhodes Jr., R. K. Chesser, and M. H. Smith (pp. 109–45). Chicago: University of Chicago Press.

Hall, A. R., and N. Colegrave. 2007. How Does Resource Supply Affect Evolutionary Diversification? *Proceedings of the Royal Society B: Biological Sciences* 274: 73–78.

Hall, A. R., V. F. Griffiths, R. C. MacLean, and N. Colegrave. 2010. Mutational Neighbourhood and Mutation Supply Rate Constrain Adaptation in *Pseudomonas aeruginosa*. *Proceedings of the Royal Society B: Biological Sciences* 277: 643–50.

Hall, A. R., J. R. Meyer, and R. Kassen. 2008. Selection for Predator Resistance Varies with Resource Supply in a Model Adaptive Radiation. *Evolutionary Ecology Research* 10: 735–46.

Hall, B. G. 1989. Selection, Adaptation, and Bacterial Operons. *Genome* 31: 265–71.

Harcombe, W. 2010. Novel Cooperation Experimentally Evolved between Species. *Evolution* 64: 2166–72.

Harmon, L. J., J. A. Schulte, A. Larson, and J. B. Losos. 2003. Tempo and Mode of Evolutionary Radiation in Iguanian Lizards. *Science* 301: 961–64.

Harrison, F. 2007. Microbial Ecology of the Cystic Fibrosis Lung. *Microbiology* 153: 917–23.

Heard, S. B., and D. L. Hauser. 1995. Key Evolutionary Innovations and Their Ecological Mechanisms. *Historical Biology* 10: 151–73.

Hedrick, P. W. 1986. Genetic Polymorphism in Heterogeneous Environments: A Decade Later. *Annual Review of Ecology and Systematics* 17: 535–66.

Helling, R. B., C. N. Vargas, and J. Adams. 1987. Evolution of *Escherichia coli* during Growth in a Constant Environment. *Genetics* 116: 349–58.

Hereford, J. 2009. A Quantitative Survey of Local Adaptation and Fitness Trade-offs. *American Naturalist* 173: 579–88.

Hermisson, J., and P. S. Pennings. 2005. Soft Sweeps: Molecular Population Genetics of Adaptation from Standing Genetic Variation. *Genetics* 169: 2335–52.

Herring, C. D., and B. Ø. Palsson. 2007. An Evaluation of Comparative Genome Sequencing (CGS) by Comparing Two Previously Sequenced Bacterial Genomes. *BMC Genomics* 8: 274.

Herring, C. D., A. Raghunathan, C. Honisch, T. Patel, M. K. Applebee, A. R. Joyce, et al. 2006. Comparative Genome Sequencing of *Escherichia coli* Allows Observation of Bacterial Evolution on a Laboratory Timescale. *Nature Genetics* 38: 1406–12.

Hietpas, R. T, J. D. Jensen, and D. N. A. Bolon. 2011. Experimental Illumination of a Fitness Landscape. *Proceedings of the National Academy of Sciences USA* 108: 7896–901.

Hillesland, K. L., and D. A. Stahl. 2010. Rapid Evolution of Stability and Productivity at the Origin of a Microbial Mutualism. *Proceedings of the National Academy of Sciences USA* 107: 2124–29.

Hoeksema, J. D., and S. E. Forde. 2008. A Meta-analysis of Factors Affecting Local Adaptation between Interacting Species. *American Naturalist* 171: 275–90.

Hoekstra, H. E., and J. A. Coyne. 2007. The Locus of Evolution: Evo Devo and the Genetics of Adaptation. *Evolution* 61: 995–1016.

Holder, K. K., and J. J. Bull. 2001. Profiles of Adaptation in Two Similar Viruses. *Genetics* 159: 1393–404.

Holt, R. D., and R. Gomulkiewicz. 1997. How Does Immigration Influence Local Adaptation? A Reexamination of a Familiar Paradigm. *American Naturalist* 149: 563–72.

Hughes, B. S., A. J. Cullum, and A. F. Bennett. 2007. An Experimental Evolutionary Study on Adaptation to Temporally Fluctuating pH in *Escherichia coli*. *Physiological and Biochemical Zoology* 80: 406–21.

Hutchinson, G. E. 1957. Concluding Remarks. *Cold Spring Harbor Symposia on Quantitative Biology* 22: 415–27.

Jacob, F. 1977. Evolution and Tinkering. *Science* 196: 1161–66.

Jacoby, G. A. 2005. Mechanisms of Resistance to Quinolones. *Clinical Infectious Diseases* 41 (Suppl. 2): S120–26.

Jasmin, J.-N., and R. Kassen. 2007a. On the Experimental Evolution of Specialization and Diversity in Heterogeneous Environments. *Ecology Letters* 10: 272–81.

———. 2007b. Evolution of a Single Niche Specialist in Variable Environments. *Proceedings of the Royal Society B: Biological Sciences* 274: 2761–67.

Jasmin, J.-N., and C. Zeyl. 2013. Evolution of Pleiotropic Costs in Experimental Populations. *Journal of Evolutionary Biology* 26: 1363–69.

Joyce, P., D. R. Rokyta, C. J. Beisel, and H. A. Orr. 2008. A General Extreme Value Theory Model for the Adaptation of DNA Sequences Under Strong Selection and Weak Mutation. *Genetics* 180: 1627–43.

Kaltz, O., and G Bell. 2002. The Ecology and Genetics of Fitness in *Chlamydomonas*. XII. Repeated Sexual Episodes Increase Rates of Adaptation to Novel Environments. *Evolution* 56: 1743–53.

Kao, K. C., and G. Sherlock. 2008. Molecular Characterization of Clonal Interference during Adaptive Evolution in Asexual Populations of *Saccharomyces cerevisiae*. *Nature Genetics* 40: 1499–504.

Kapranov, P., and G. St. Laurent. 2012. Dark Matter RNA: Existence, Function, and Controversy. *Frontiers in Genetics* 3: 60.

Kassen, R. 2002. The Experimental Evolution of Specialists, Generalists, and the Maintenance of Diversity. *Journal of Evolutionary Biology* 15: 173–90.

Kassen, R., and T. Bataillon. 2006. Distribution of Fitness Effects among Beneficial Mutations before Selection in Experimental Populations of Bacteria. *Nature Genetics* 38: 484–88.

Kassen, R., and G. Bell. 1998. Experimental Evolution in *Chlamydomonas*. IV. Selection in Environments That Vary through Time at Different Scales. *Heredity* 80: 732–41.

———. 2000. The Ecology and Genetics of Fitness in *Chlamydomonas*. X. The Relationship between Genetic Correlation and Genetic Distance. *Evolution* 54: 425–32.

Kassen, R., A. Buckling, G. Bell, and P. B. Rainey. 2000. Diversity Peaks at Intermediate Productivity in a Laboratory Microcosm. *Nature* 406: 508–12.

Kassen, R., M. Llewellyn, and P. B. Rainey. 2004. Ecological Constraints on Diversification in a Model Adaptive Radiation. *Nature* 431: 984–88.

Kawecki, T. J., and D. Ebert. 2004. Conceptual Issues in Local Adaptation. *Ecology Letters* 7: 1225–41.

Kent, R. J., L. C. Harrington, and D. E. Norris. 2007. Genetic Differences between *Culex pipiens f. molestus* and *Culex pipiens pipiens* (Diptera: Culicidae) in New York. *Journal of Medical Entomology* 44: 50–59.

Kerr, B., M. A. Riley, M. W. Feldman, and B. J. M. Bohannan. 2002. Local Dispersal Promotes Biodiversity in a Real-Life Game of Rock-paper-scissors. *Nature* 418: 171–74.

Khan, A. I., D. M. Dinh, D. Schneider, R. E. Lenski, and T. F. Cooper. 2011. Negative Epistasis between Beneficial Mutations in an Evolving Bacterial Population. *Science* 332: 1193–96.

Kinnersley, M., W. E. Holben, and F. Rosenzweig. 2009. E Unibus Plurum: Genomic Analysis of an Experimentally Evolved Polymorphism in *Escherichia coli*. *PLoS Genetics* 5: e1000713.

Knies, J. L., R. Izem, K. L. Supler, J. G. Kingsolver, and C. L. Burch. 2006. The Genetic Basis of Thermal Reaction Norm Evolution in Lab and Natural Phage Populations. *PLoS Biology* 4: e201.

Knight, C. G., N. Zitzmann, S. Prabhakar, R. Antrobus, R. Dwek, H. Hebestreit, et al. 2006. Unraveling Adaptive Evolution: How a Single Point Mutation Affects the Protein Coregulation Network. *Nature Genetics* 38: 1015–22.

Kopp, M., and J. Hermisson. 2009. The Genetic Basis of Phenotypic Adaptation II: The Distribution of Adaptive Substitutions in the Moving Optimum Model. *Genetics* 183: 1453–76.

Korona, R. 1996. Adaptation to Structurally Different Environments. *Proceedings of the Royal Society B:Biological Sciences* 263: 1665–69.

Korona, R., C. H. Nakatsu, L. J. Forney, and R. E. Lenski. 1994. Evidence for Multiple Adaptive Peaks from Populations of Bacteria Evolving in a Structured Habitat. *Proceedings of the National Academy of Sciences USA* 91: 9037–41.

Koskella, B. 2013. Phage-Mediated Selection on Microbiota of a Long-Lived Host. *Current Biology* 23: 1256–60.

Kothera, L., M. Godsey, J.-P. Mutebi, and H. M. Savage. 2010. A Comparison of Aboveground and Belowground Populations of *Culex pipiens* (Diptera: Culicidae) Mosquitoes in Chicago, Illinois, and New York City, New York, Using Microsatellites. *Journal of Medical Entomology* 47: 805–13.

Kryazhimskiy, S., D. P. Rice, and M. M. Desai. 2012. Population Subdivision and Adaptation in Asexual Populations of *Saccharomyces cerevisiae*. *Evolution* 66: 1931–41.

Kvitek, D. J., and G. Sherlock. 2011. Reciprocal Sign Epistasis between Frequently Experimentally Evolved Adaptive Mutations Causes a Rugged Fitness Landscape. *PLoS Genetics* 7: e1002056.

Lang, G. I., D. Botstein, and M. M. Desai. 2011. Genetic Variation and the Fate of Beneficial Mutations in Asexual Populations. *Genetics* 188: 647–61.

Lawrie, D. S., P. W. Messer, R. Hershberg, and D. A. Petrov. 2013. Strong Purifying Selection at Synonymous Sites in *D. melanogaster*. *PLoS Genetics* 9: e1003527.

Lee, D.-H., and B. Ø. Palsson. 2010. Adaptive Evolution of *Escherichia coli* K-12 MG1655 during Growth on a Nonnative Carbon Source, L-1,2-propanediol. *Applied and Environmental Microbiology* 76: 4158–68.

Lee, M.-C., H.-H. Chou, and C. J. Marx. 2009. Asymmetric, Bimodal Trade-offs during Adaptation of Methylobacterium to Distinct Growth Substrates. *Evolution* 63: 2816–30.

Lenski, R. E., and B. R. Levin. 1985. Constraints on the Coevolution of Bacteria and Virulent Phage: A Model, Some Experiments, and Predictions for Natural Communities. *American Naturalist* 125: 585–602.

Lenski, R. E., M. R. Rose, S. C. Simpson, and S. C. Tadler. 1991. Long-Term Experimental Evolution in *Escherichia coli*. I. Adaptation and Divergence during 2,000 Generations. *American Naturalist* 138: 1315–41.

Lenski, R. E., and M. Travisano. 1994. Dynamics of Adaptation and Diversification: A 10,000-Generation Experiment with Bacterial Populations. *Proceedings of the National Academy of Sciences USA* 91: 6808–14.

Levene, H. 1953. Genetic Equilibrium When More than One Ecological Niche Is Available. *American Naturalist* 87: 331–33.

Levins, R. 1968. *Evolution in Changing Environments*. Princeton, NJ: Princeton University Press.

Lewontin, R. C. 1974. *The Genetic Basis of Evolutionary Change*. New York: Columbia University Press.

Lin, E. C. C. 1976. Glycerol Dissimilation and Its Regulation in Bacteria. *Annual Review of Microbiology* 30: 535–78.

Lin, E. C. C., A. J. Hacking, and J. Aguilar. 1976. Experimental Models of Acquisitive Evolution. *BioScience* 26: 548–55.

Lohse, K., A. Gutierrez, and O. Kaltz. 2006. Experimental Evolution of Resistance in *Paramecium caudatum* Against the Bacterial Parasite *Holospora undulata*. *Evolution* 60: 1177–86.

Long, M. 2001. Evolution of Novel Genes. *Current Opinion in Genetics & Development* 11: 673–80.

Losos, J. B. 2010. Adaptive Radiation, Ecological Opportunity, and Evolutionary Determinism. *American Naturalist* 175: 623–39.

———. 2011. Seeing the Forest for the Trees: The Limitations of Phylogenies in Comparative Biology. *American Naturalist* 177: 709–27.

Losos, J. B., and R. E. Ricklefs. 2009. Adaptation and Diversification on Islands. *Nature* 457: 830–36.

Lynch, M. 2010. Evolution of the Mutation Rate. *Trends in Genetics* 26: 345–52.

Lynch, M., and B. Walsh. 1997. *Genetics and Analysis of Quantitative Traits*. Sunderland, MA: Sinauer Associates.

MacLean, R. C., and G. Bell. 2002. Experimental Adaptive Radiation in *Pseudomonas*. *American Naturalist* 160: 569–81.

MacLean, R. C., G. Bell, and P. B. Rainey. 2004. The Evolution of a Pleiotropic Fitness Tradeoff in *Pseudomonas fluorescens*. *Proceedings of the National Academy of Sciences USA* 101: 8072–77.

MacLean, R. C., and A. Buckling. 2009. The Distribution of Fitness Effects of Beneficial Mutations in *Pseudomonas aeruginosa*. *PLoS Genetics* 5: e1000406.

MacLean, R. C., A. Dickson, and G. Bell. 2005. Resource Competition and Adaptive Radiation in a Microbial Microcosm. *Ecology Letters* 8: 38–46.

MacLean, R. C., G. G. Perron, and A. Gardner. 2010. Diminishing Returns from Beneficial Mutations and Pervasive Epistasis Shape the Fitness Landscape for Rifampicin Resistance in *Pseudomonas aeruginosa. Genetics* 186: 1345–54.

Maharjan, R., S. Seeto, L. Notley-McRobb, and T. Ferenci. 2006. Clonal Adaptive Radiation in a Constant Environment. *Science* 313: 514–17.

Maharjan, R., Z. Zhou, Y. Ren, Y. Li, J. Gaffé, D. Schneider, et al. 2010. Genomic Identification of a Novel Mutation in *hfq* That Provides Multiple Benefits in Evolving Glucose-Limited Populations of *Escherichia coli. Journal of Bacteriology* 192: 4517–21.

Mank, J. E. 2007. Mating Preferences, Sexual Selection and Patterns of Cladogenesis in Ray-finned Fishes. *Journal of Evolutionary Biology* 20 (2): 597–602.

Margulis, L., and R. Fester (eds.). 1991. *Symbiosis as a Source of Evolutionary Innovation.* Boston: MIT Press.

Marston, M. F., F. J. Pierciey, A. Shepard, G. Gearin, J. Qi, C. Yandava, et al. 2012. Rapid Diversification of Coevolving Marine Synechococcus and a Virus. *Proceedings of the National Academy of Sciences USA* 109: 4544–49.

Martin, G., and T. Lenormand. 2008. The Distribution of Beneficial and Fixed Mutation Fitness Effects Close to an Optimum. *Genetics* 179: 907–16.

Massin, N., and A. Gonzalez. 2006. Adaptive Radiation in a Fluctuating Environment: Disturbance Affects the Evolution of Diversity in a Bacterial Microcosm. *Evolutionary Ecology Research* 86: 2815–24.

Mattingly, P. F., L. E. Rozeboom, K. L. Knight, H. Laven, F. H. Drummond, S. R. Christophers, et al. 1951. The *Culex pipiens* Complex. *Transactions of the Royal Entomological Society of London* 102: 331–42.

Maughan, H., V. Callicotte, A. Hancock, C. W. Birky, W. L. Nicholson, and J. Masel. 2006. The Population Genetics of Phenotypic Deterioration in Experimental Populations of *Bacillus subtilis. Evolution* 60: 686–95.

Maynard Smith, J. 1989. The Causes of Extinction. *Philosophical Transactions of the Royal Society B: Biological Sciences* 325: 241–52.

Mayr, E. 1963. *Animal Species and Evolution.* Cambridge, MA: Harvard University Press.

McDonald, M. J., T. F. Cooper, H. J. E. Beaumont, and P. B. Rainey. 2011. The Distribution of Fitness Effects of New Beneficial Mutations in *Pseudomonas fluorescens. Biology Letters* 7: 98–100.

McDonald, M. J., S. M. Gehrig, P. L. Meintjes, X.-X. Zhang, and P. B. Rainey. 2009. Adaptive Divergence in Experimental Populations of *Pseudomonas fluorescens.* IV. Genetic Constraints Guide Evolutionary Trajectories in a Parallel Adaptive Radiation. *Genetics* 183: 1041–53.

McGrady-Steed, J., and P. J. Morin. 2000. Biodiversity, Density Compensation, and the Dynamics of Populations and Functional Groups. *Ecology* 81: 361–73.

McPeek, M. 2008. The Ecological Dynamics of Clade Diversification and Community Assembly. *American Naturalist* 172: E270–84.

Melnyk, A., and R. Kassen. 2011. Adaptive Landscapes in Evolving Populations of *Pseudomonas fluorescens. Evolution* 65: 3048–59.

Meyer, A. 1993. Phylogenetic Relationships and Evolutionary Processes in East African Cichlid Fishes. *Trends in Ecology and Evolution* 8: 279–84.

Meyer, J. R., D. T. Dobias, J. S. Weitz, J. E. Barrick, R. T. Quick, and R. E. Lenski. 2012. Repeatability and Contingency in the Evolution of a Key Innovation in Phage Lambda. *Science* 335: 428–32.

Meyer, J. R., and R. Kassen. 2007. The Effects of Competition and Predation on Diversification in a Model Adaptive Radiation. *Nature* 446: 432–35.

Meyer, J. R., S. E. Schoustra, J. Lachapelle, and R. Kassen. 2011. Overshooting Dynamics in a Model Adaptive Radiation. *Proceedings of the Royal Society B: Biological Sciences* 278: 392–98.

Miller, C. R., P. Joyce, and H. A. Wichman. 2011. Mutational Effects and Population Dynamics during Viral Adaptation Challenge Current Models. *Genetics* 187: 185–202.

Minty, J. J., A. A. Lesnefsky, F. Lin, Y. Chen, T. A. Zaroff, A. B. Veloso, et al. 2011. Evolution Combined with Genomic Study Elucidates Genetic Bases of Isobutanol Tolerance in *Escherichia coli*. *Microbial Cell Factories* 10: 18.

Miralles, R., A. Moya, and S. F. Elena. 2000. Diminishing Returns of Population Size in the Rate of RNA Virus Adaptation. *Journal of Virology* 74: 3566–71.

Mitter, C., B. Farrell, and B. Wiegmann. 1988. Phylogenetic Study of Adaptive Zones: Has Phytophagy Promoted Insect Diversification? *American Naturalist* 132: 107–28.

Mongold, J. A., A. F. Bennett, and R. E. Lenski. 1996. Evolutionary Adaptation to Temperature. IV. Adaptation of *Escherichia coli* at a Niche Boundary. *Evolution* 50: 35–43.

Monod. J. 1949. The Growth of Bacterial Cultures. *Annual Review of Microbiology* 3: 371–94.

Morin, P. J. 2011. *Community Ecology* (2nd ed.). Oxford: Wiley-Blackwell.

Morlon, H., B. D. Kemps, J. B. Plotkin, and D. Brisson. 2012. Explosive Radiation of a Bacterial Species Group. *Evolution* 66: 2577–86.

Mortlock, R. P. 1983. Experiments in Evolution Using Microorganisms. *BioScience* 33: 308–13.

Mousseau, T. A., and D. A. Roff. 1987. Natural Selection and the Heritability of Fitness Components. *Heredity* 59: 181–97.

Murray, G. M., and J. P. Brennan. 2009. Estimating Disease Losses to the Australian Wheat Industry. *Australasian Plant Pathology* 38: 558–70.

Nagylaki, T., and Y. Lou. 2006. Evolution Under the Multiallelic Levene Model. *Theoretical Population Biology* 70: 401–11.

Nee, S. 2006. Birth-Death Models in Macroevolution. *Annual Review of Ecology, Evolution, and Systematics* 37: 1–17.

Nosil, P. 2012. *Ecological Speciation*. Oxford: Oxford University Press.

Nosil, P., and J. L. Feder. 2012. Genomic Divergence during Speciation: Causes and Consequences. *Philosophical Transactions of the Royal Society B: Biological Sciences* 367: 332–42.

Nosil, P., D. J. Funk, and D. Ortiz-Barrientos. 2009. Divergent Selection and Heterogeneous Genomic Divergence. *Molecular Ecology* 18: 375–402.

Nosil, P., and L. J. Harmon. 2009. Niche Dimensionality and Ecological Speciation. In *Speciation and Patterns of Diversity,* eds. R. Butlin, J. Bridle, and D. Schluter. Cambridge: Cambridge University Press.

Nosil, P., and C. P. Sandoval. 2008. Ecological Niche Dimensionality and the Evolutionary Diversification of Stick Insects. *PloS One* 3: e1907.

Notley-McRobb, L., and T. Ferenci. 1999a. Adaptive *mgl*-Regulatory Mutations and Genetic Diversity Evolving in Glucose-Limited *Escherichia coli* Populations. *Environmental Microbiology* 1: 33–43.

———. 1999b. The Generation of Multiple Co-existing *mal*-Regulatory Mutations through Polygenic Evolution in Glucose-Limited Populations of *Escherichia coli*. *Environmental Microbiology* 1: 45–52.

Novella, I. S., S. Zarate, D. Metzgar, and B. E. Ebendick-Corpus. 2004. Positive Selection of Synonymous Mutations in Vesicular Stomatitis Virus. *Journal of Molecular Biology* 342: 1415–21.

Novick, A., and L. Szilard. 1950. Experiments with the Chemostat on Spontaneous Mutations of Bacteria. *Proceedings of the National Academy of Sciences USA* 36: 708–19.

Odling-Smee, F. J., K. N. Laland, and M. W. Feldman. 2003. *Niche Construction: The Neglected Process in Evolution*. Princeton, NJ: Princeton University Press.

Ohno, S. 1970. *Evolution by Gene Duplication*. New York: Springer Verlag.

Orr, H. A. 2002. The Population Genetics of Adaptation: The Adaptation of DNA Sequences. *Evolution* 56: 1317–30.

———. 2005. The Probability of Parallel Evolution. *Evolution* 59: 216–20.

Orr, H. A., and S. P. Otto. 1994. Does Diploidy Increase the Rate of Adaptation? *Genetics* 136: 1475–80.

Orr, H. A., and R. L. Unckless. 2008. Population Extinction and the Genetics of Adaptation. *The American Naturalist* 172: 160–69.

Ostrowski, E., D. E. Rozen, and R. E. Lenski. 2005. Pleiotropic Effects of Beneficial Mutations in *Escherichia coli*. *Evolution* 59: 2343–52.

Ostrowski, E. A., R. J. Woods, and R. E. Lenski. 2008. The Genetic Basis of Parallel and Divergent Phenotypic Responses in Evolving Populations of *Escherichia coli*. *Proceedings of the Royal Society B: Biological Sciences* 275: 277–84.

Otto, S. P. 2007. The Evolutionary Consequences of Polyploidy. *Cell* 131: 452–62.

Otto, S. P., and C. D. Jones. 2000. Detecting the Undetected: Estimating the Total Number of Loci Underlying a Quantitative Trait. *Genetics* 156: 2093–107.

Paquin, C. E., and J. Adams. 1983. Relative Fitness Can Decrease in Evolving Asexual Populations of *S. cerevisiae*. *Nature* 306: 368–71.

Parmley, J. L., and L. D. Hurst. 2007. How Do Synonymous Mutations Affect Fitness? *BioEssays* 29: 515–19.

Paterson, S., T. Vogwill, A. Buckling, R. Benmayor, A. J. Spiers, N. R. Thomson, et al. 2010. Antagonistic Coevolution Accelerates Molecular Evolution. *Nature* 464: 275–78.

Perfeito, L., L. Fernandes, C. Mota, and I. Gordo. 2007. Adaptive Mutations in Bacteria: High Rate and Small Effects. *Science* 317: 813–15.

Perfeito, L., M. I. Pereira, P. R. Campos, and I. Gordo. 2008. The Effect of Spatial Structure on Adaptation in *Escherichia coli*. *Biology Letters* 4: 57–59.

Perron, G. G., A. Gonzalez, and A. Buckling. 2007. Source-Sink Dynamics Shape the Evolution of Antibiotic Resistance and Its Pleiotropic Fitness Cost. *Proceedings of the Royal Society B: Biological Sciences* 274: 2351–56.

———. 2008. The Rate of Environmental Change Drives Adaptation to an Antibiotic Sink. *Journal of Evolutionary Biology* 21: 1724–31.

Pinto, G., D. L. Mahler, L. J. Harmon, and J. B. Losos. 2008. Testing the Island Effect in Adaptive Radiation: Rates and Patterns of Morphological Diversification in Caribbean and Mainland Anolis Lizards. *Proceedings of the Royal Society B: Biological Sciences* 275: 2749–57.

Plotkin, J. B., and G. Kudla. 2011. Synonymous but Not the Same: The Causes and Consequences of Codon Bias. *Nature Reviews Genetics* 12: 32–42.

Poltak, S. R., and V. S. Cooper. 2011. Ecological Succession in Long-Term Experimentally Evolved Biofilms Produces Synergistic Communities. *ISME Journal* 5: 369–78.

Ponciano, J. M., H.-J. La, P. Joyce, and L. J. Forney. 2009. Evolution of Diversity in Spatially Structured *Escherichia coli* Populations. *Applied and Environmental Microbiology* 75: 6047–54.

Poole, K. 2005. Efflux-Mediated Antimicrobial Resistance. *Journal of Antimicrobial Chemotherapy* 56: 20–51.

Pränting, M., and D. I. Andersson. 2011. Escape from Growth Restriction in Small Colony Variants of *Salmonella typhimurium* by Gene Amplification and Mutation. *Molecular Microbiology* 79: 305–15.

Presgraves, D. C. 2010. The Molecular Evolutionary Basis of Species Formation. *Nature Reviews Genetics* 11: 175–80.

Presloid, J. B., B. E. Ebendick-Corpus, S. Zárate, and I. S. Novella. 2008. Antagonistic Pleiotropy Involving Promoter Sequences in a Virus. *Journal of Molecular Biology* 382: 342–52.

Primmer, C. R. 2011. Genetics of Local Adaptation in Salmonid Fishes. *Heredity* 106: 401–3.

Rabosky, D. L., and I. J. Lovette. 2008. Density-Dependent Diversification in North American Wood Warblers. *Proceedings of the Royal Society B: Biological Sciences* 275: 2363–71.

Rainey, P. B., and K. Rainey. 2003. Evolution of Cooperation and Conflict in Experimental Bacterial Populations. *Nature* 425: 72–74.

Rainey, P. B., and M. Travisano. 1998. Adaptive Radiation in a Heterogeneous Environment. *Nature* 394: 69–72.

Raup, D. M., S. J. Gould, T. J. M. Schopf, and D. S. Simberloff. 1973. Stochastic Models of Phylogeny and the Evolution of Diversity. *Journal of Geology* 81: 525–42.

Reboud, X., and G. Bell. 1997. Experimental Evolution in *Chlamydomonas*. III. Evolution of Specialist and Generalist Types in Environments That Vary in Space and Time. *Heredity* 78: 507–14.

Remold, S. K., A. Rambaut, and P. E. Turner. 2008. Evolutionary Genomics of Host Adaptation in Vesicular Stomatitis Virus. *Molecular Biology and Evolution* 25: 1138–47.

Reusken, C., A. De Vries, J. Buijs, M. A. H. Braks, W. den Hartog, and E. J. Scholte. 2010. First Evidence for Presence of *Culex pipiens* Biotype *molestus* in the Netherlands, and of Hybrid Biotype *pipiens* and *molestus* in Northern Europe. *Journal of Vector Ecology* 35: 210–12.

Reznick, D. N., and C. K. Ghalambor. 2001. The Population Ecology of Contemporary Adaptations: What Empirical Studies Reveal about the Conditions That Promote Adaptive Evolution. *Genetica* 112–113: 183–98.

Rice, W. R., and E. E. Hostert. 1993. Laboratory Experiments on Speciation: What Have We Learned in 40 Years? *Evolution* 47: 1637–53.

Ricklefs, R. E. 2006. Evolutionary Diversification and the Origin of the Diversity-Environment Relationship. *Ecology* 87: S3–13.

———. 2007. Estimating Diversification Rates from Phylogenetic Information. *Trends in Ecology and Evolution* 22: 601–10.

Ritchie, M. G. 2007. Sexual Selection and Speciation. *Annual Review of Ecology, Evolution, and Systematics* 38: 79–102.

Robertson, A. 1959. The Sampling Variance of the Genetic Correlation Coefficient. *Biometrics* 15: 469–85.

Rockman, M. V. 2012. The QTN Program and the Alleles That Matter for Evolution: All That's Gold Does Not Glitter. *Evolution* 66: 1–17.

Rogers, D. W., and D. Greig. 2009. Experimental Evolution of a Sexually Selected Display in Yeast. *Proceedings of the Royal Society B: Biological Sciences* 276: 543–49.

Rokyta, D. R., C. J. Beisel, P. Joyce, M. T. Ferris, C. L. Burch, and H. A. Wichman. 2008. Beneficial Fitness Effects Are Not Exponential for Two Viruses. *Journal of Molecular Evolution* 67: 368–76.

Rokyta, D. R., Z. Abdo, and H. A. Wichman. 2009. The Genetics of Adaptation for Eight Microvirid Bacteriophages. *Journal of Molecular Evolution* 69: 229–39.

Rokyta, D. R., P. Joyce, S. B. Caudle, and H. A. Wichman. 2005. An Empirical Test of the Mutational Landscape Model of Adaptation Using a Single-Stranded DNA Virus. *Nature Genetics* 37: 441–44.

Rosenzweig, M. L. 1995. *Species Diversity in Space and Time.* Cambridge: Cambridge University Press.

Rosenzweig, R. F., R. R. Sharp, D. S. Treves, and J. Adams. 1994. Microbial Evolution in a Simple Unstructured Environment: Genetic Differentiation in *Escherichia coli*. *Genetics* 137: 903–17.

Royer, D. L., L. J. Hickey, and S. L. Wing. 2003. Ecological Conservatism in the "Living Fossil" Gingko. *Paleobiology* 29: 84–104.

Rozen, D. E., J. A. G. M. de Visser, and P. J. Gerrish. 2002. Fitness Effects of Fixed Beneficial Mutations in Microbial Populations. *Current Biology* 12: 1040–45.

Rozen, D. E., M. G. J. L. Habets, A. Handel, and J. A. G. M. de Visser. 2008. Heterogeneous Adaptive Trajectories of Small Populations on Complex Fitness Landscapes. *PloS One* 3: e1715.

Rozen, D. E., and R. E. Lenski. 2000. Long-Term Experimental Evolution in *Escherichia coli*. VIII. Dynamics of a Balanced Polymorphism. *American Naturalist* 155: 24–35.

Rozen, D. E., N. Philippe, J. A. G. M. de Visser, R. E. Lenski, and D. Schneider. 2009. Death and Cannibalism in a Seasonal Environment Facilitate Bacterial Coexistence. *Ecology Letters* 12: 34–44.

Rueffler, C., T. J. M. Van Dooren, O. Leimar, and P. A. Abrams. 2006. Disruptive Selection and Then What? *Trends in Ecology and Evolution* 21: 238–45.

Rundle, H. D., S. F. Chenoweth, and M. W. Blows. 2006. The Roles of Natural and Sexual Selection during Adaptation to a Novel Environment. *Evolution* 60: 2218–25.

Rundle, H. D., and P. Nosil. 2005. Ecological Speciation. *Ecology Letters* 8: 336–52.

Salverda, M. L. M., E. Dellus, F. A. Gorter, A. J. M. Debets, J. van der Oost, R. F. Hoekstra, et al. 2011. Initial Mutations Direct Alternative Pathways of Protein Evolution. *PLoS Genetics* 7: e1001321.

Sanford, E., and M. W. Kelly. 2011. Local Adaptation in Marine Invertebrates. *Annual Review of Marine Science* 3: 509–35.

Sanjuán, R., A. Moya, and S. F. Elena. 2004. The Distribution of Fitness Effects Caused by Single-Nucleotide Substitutions in an RNA Virus. *Proceedings of the National Academy of Sciences USA* 101: 8396–401.

Scheiner, S. M., and L. Y. Yampolsky. 1998. The Evolution of *Daphnia pulex* in a Temporally Varying Environment. *Genetical Research* 72: 25–37.

Schluter, D. 2000a. *The Ecology of Adaptive Radiation*. Oxford: Oxford University Press.

———. 2000b. Ecological Character Displacement in Adaptive Radiation. *American Naturalist* 156 (Suppl): S4–16.

———. 2009. Evidence for Ecological Speciation and Its Alternative. *Science* 323: 737–41.

Schluter, D., K. B. Marchinko, R. D. H. Barrett, and S. M. Rogers. 2010. Natural Selection and the Genetics of Adaptation in Threespine Stickleback. *Philosophical Transactions of the Royal Society B: Biological Sciences* 365: 2479–86.

Schoustra, S. E., T. Bataillon, D. R. Gifford, and R. Kassen. 2009. The Properties of Adaptive Walks in Evolving Populations of Fungus. *PLoS Biology* 7: e1000250.

Schoustra, S. E., A. J. M. Debets, M. Slakhorst, and R. F. Hoekstra. 2006. Reducing the Cost of Resistance; Experimental Evolution in the Filamentous Fungus *Aspergillus nidulans*. *Journal of Evolutionary Biology* 19: 1115–27.

———. 2007. Mitotic Recombination Accelerates Adaptation in the Fungus *Aspergillus nidulans*. *PLoS Genetics* 3: e68.

Schoustra, S. E., C. Kasase, C. Toarta, R. Kassen, and A. J. Poulain. 2013. Microbial Community Structure of Three Traditional Zambian Fermented Products: Mabisi, Chibwantu and Munkoyo. *PloS One* 8: e63948.

Schoustra, S. E., D. Punzalan, R. Dali, H.D. Rundle, and R. Kassen. 2012. Multivariate Phenotypic Divergene Due to the Fixation of Beneficial Mutations in Experimentally Evolved Lineages of a Filamentous Fungus. *PLoS ONE* 7: e50305.

Schoustra, S., H. D. Rundle, R. Dali, and R. Kassen. 2010. Fitness-Associated Sexual Reproduction in a Filamentous Fungus. *Current Biology* 20: 1350–55.

Schrag, S. J., and J. E. Mittler. 1996. Host-Parasite Coexistence: The Role of Spatial Refuges in Stabilizing Bacteria-Phage Interactions. *American Naturalist* 148: 348–77.

Schulte, R. D., C. Makus, B. Hasert, N. K. Michiels, and H. Schulenburg. 2010. Multiple Reciprocal Adaptations and Rapid Genetic Change upon Experimental Coevolution of an Animal Host and Its Microbial Parasite. *Proceedings of the National Academy of Sciences USA* 107: 7359–64.

Sepkoski, J. J. 1978. A Kinetic Model of Phanerozoic Taxonomic Diversity I. Analysis of Marine Orders. *Paleobiology* 4: 223–51.

————. 1998. Rates of Speciation in the Fossil Record. *Philosophical Transactions of the Royal Society B: Biological Sciences* 353: 315–26.

Shapiro, B. J., J. Friedman, O. X. Cordero, S. P. Preheim, S. C. Timberlake, G. Szabó, et al. 2012. Population Genomics of Early Events in the Ecological Differentiation of Bacteria. *Science* 336: 48–51.

Shendure, J., G. J. Porreca, N. B. Reppas, X. Lin, J. P. McCutcheon, A. M. Rosenbaum, et al. 2005. Accurate Multiplex Polony Sequencing of an Evolved Bacterial Genome. *Science* 309: 1728–32.

Shute, P. G. 1951. *Culex molestus. Transactions of the Royal Entomological Society of London* 102: 380–82.

Silvertown, J., P. Poulton, E. Johnston, G. Edwards, M. Heard, and P. M. Biss. 2006. The Park Grass Experiment 1856–2006: Its Contribution to Ecology. *Journal of Ecology* 94: 801–14.

Simpson, G. G. 1953. *The Major Features of Evolution.* New York: Columbia University Press.

Smith, D. R., A. R. Quinlan, H. E. Peckham, K. Makowsky, W. Tao, B. Woolf, et al. 2008. Rapid Whole-genome Mutational Profiling using Next-generation Sequencing Technologies. *Genome Research* 18: 1638–42.

Smith, E. E., D. G. Buckley, Z. Wu, C. Saenphimmachak, L. R. Hoffman, D. A. D'Argenio, et al. 2006. Genetic Adaptation by *Pseudomonas aeruginosa* to the Airways of Cystic Fibrosis Patients. *Proceedings of the National Academy of Sciences USA* 103: 8487–92.

Smith-Tsurkan, S. D., C. O. Wilke, and I. S. Novella. 2010. Incongruent Fitness Landscapes, Not Tradeoffs, Dominate the Adaptation of Vesicular Stomatitis Virus to Novel Host Types. *Journal of General Virology* 91: 1484–93.

Sniegowski, P. D., P. J. Gerrish, and R. E. Lenski. 1997. Evolution of High Mutation Rates in Experimental Populations of *Escherichia coli. Nature* 387: 703–5.

Sniegowski, P. D., P. J. Gerrish, T. Johnson, and A. Shaver. 2000. The Evolution of Mutation Rates: Separating Causes from Consequences. *BioEssays* 22: 1057–66.

Sommer, U. 1985. Comparison between Steady State and Non-Steady State Competition: Experiments with Natural Phytoplankton. *Limnology and Oceanography* 30: 335–46.

Spencer, C. C., M. Bertrand, M. Travisano, and M. Doebeli. 2007. Adaptive Diversification in Genes That Regulate Resource Use in *Escherichia coli. PLoS Genetics* 3: e15.

Spencer, C. C., J. Tyerman, M. Bertrand, and M. Doebeli. 2008. Adaptation Increases the Likelihood of Diversification in an Experimental Bacterial Lineage. *Proceedings of the National Academy of Sciences USA* 105: 1585–89.

Stanley, S. M. 2007. An Analysis of the History of Marine Animal Diversity. *Paleobiology* 33: 1–55.

Stern, D. L., and V. Orgogozo. 2009. Is Genetic Evolution Predictable? *Science* 323: 746–51.

Stevens, M. H. H., M. Sanchez, J. Lee, and S. E. Finkel. 2007. Diversification Rates Increase with Population Size and Resource Concentration in an Unstructured Habitat. *Genetics* 177: 2243–50.

Stoebel, D. M., A. M. Dean, and D. E. Dykhuizen. 2008. The Cost of Expression of *Escherichia coli lac* Operon Proteins Is in the Process, Not in the Products. *Genetics* 178: 1653–60.

Suiter, A. M., O. Bänziger, and A. M. Dean. 2003. Fitness Consequences of a Regulatory Polymorphism in a Seasonal Environment. *Proceedings of the National Academy of Sciences USA* 100: 12782–86.

Summers, W. C. 1999. *Felix d'Herelle and the Origins of Molecular Biology.* New Haven, CT: Yale University Press.

Sun, S., O. G. Berg, J. R. Roth, and D. I. Andersson. 2009. Contribution of Gene Amplification to Evolution of Increased Antibiotic Resistance in *Salmonella typhimurium. Genetics* 182: 1183–95.

Tenaillon, O., A. Rodríguez-Verdugo, R. L. Gaut, P. McDonald, A. F. Bennett, A. D. Long, et al. 2012. The Molecular Diversity of Adaptive Convergence. *Science* 335: 457–61.

Teotónio, H., I. M. Chelo, M. Bradić, M. R. Rose, and A. D. Long. 2009. Experimental Evolution Reveals Natural Selection on Standing Genetic Variation. *Nature Genetics* 41: 251–57.

Thompson, D., M. M. Desai, and A. W. Murray. 2006. Ploidy Controls the Success of Mutators and Nature of Mutations during Budding Yeast Evolution. *Current Biology* 16: 1581–90.

Tilman, D. 1977. Resource Competition between Plankton Algae: An Experimental and Theoretical Approach. *Ecology* 58: 338–48.

———. 1982. *Resource Competition and Community Structure.* Princeton, NJ: Princeton University Press.

Travisano, M., and R. E. Lenski. 1996. Long-Term Experimental Evolution in *Escherichia coli.* IV. Targets of Selection and the Specificity of Adaptation. *Genetics* 143: 15–26.

Travisano, M., J. A. Mongold, A. F. Bennett, and R. E. Lenski. 1995. Experimental Tests of the Roles of Adaptation, Chance, and History in Evolution. *Science* 267: 87–90.

Treves, D. S., S. Manning, and J. Adams. 1998. Repeated Evolution of an Acetate-Crossfeeding Polymorphism in Long-Term Populations of *Escherichia coli. Molecular Biology and Evolution* 15: 789–97.

Turner, P. E. and S. F. Elena. 2000. Cost of Host Radiation in an RNA Virus. *Genetics* 156: 1465–70.

Turner, P. E., V. Souza, and R. E. Lenski. 1996. Tests of Ecological Mechanisms Promoting the Stable Coexistence of Two Bacterial Genotypes. *Ecology* 77: 2119–29.

Turner, T. L., M. W. Hahn, and S. V. Nuzhdin. 2005. Genomic Islands of Speciation in *Anopheles gambiae. PLoS Biology* 3: e285.

Tyerman, J. G., M. Bertrand, C. C. Spencer, and M. Doebeli. 2008. Experimental Demonstration of Ecological Character Displacement. *BMC Evolutionary Biology* 8: 34.

Vamosi, S. M. 2005. On the Role of Enemies in Divergence and Diversification of Prey: A Review and Synthesis. *Canadian Journal of Zoology* 83: 894–910.

Van Valen, L. 1973. A New Evolutionary Law. *Evolutionary Theory* 1: 1–30.

Vasilakis, N., E. R. Deardorff, J. L. Kenney, S. L. Rossi, K. A. Hanley, and S. C. Weaver. 2009. Mosquitoes Put the Brake on Arbovirus Evolution: Experimental Evolution Reveals Slower Mutation Accumulation in Mosquito than Vertebrate Cells. *PLoS Pathogens* 5: e1000467.

Velicer, G. J., G. Raddatz, H. Keller, S. Deiss, C. Lanz, I. Dinkelacker, et al. 2006. Comprehensive Mutation Identification in an Evolved Bacterial Cooperator and Its Cheating Ancestor. *Proceedings of the National Academy of Sciences USA* 103: 8107–12.

Vellend, M., and M. A. Geber. 2005. Connections between Species Diversity and Genetic Diversity. *Ecology Letters* 8: 767–81.

Venail, P. A., O. Kaltz, I. Olivieri, T. Pommier, and N. Mouquet. 2011. Diversification in Temporally Heterogeneous Environments: Effect of the Grain in Experimental Bacterial Populations. *Journal of Evolutionary Biology* 24: 2485–95.

Venail, P. A., R. C. MacLean, T. Bouvier, M. A. Brockhurst, M. E. Hochberg, and N. Mouquet. 2008. Diversity and Productivity Peak at Intermediate Dispersal Rate in Evolving Metacommunities. *Nature* 452: 210–14.

Vermeij, G. J. 1994. The Evolutionary Interaction among Species: Selection, Escalation, and Coevolution. *Annual Review of Ecology and Systematics* 25: 219–36.

Via, S., R. Gomulkiewicz, G. de Jong, S. M. Scheiner, C. D. Schlichting, and P. H. van Tienderen. 1995. Adaptive Phenotypic Plasticity: Consensus and Controversy. *Trends in Ecology and Evolution* 10: 212–17.

Via, S., and R. Lande. 1985. Genotype-Environment Interaction and the Evolution of Phenotypic Plasticity. *Evolution* 39: 505–22.

Villanueva, B., and B. W. Kennedy. 1992. Asymmetrical Correlated Responses to Selection Under an Infinitesimal Genetic Model. *Theoretical and Applied Genetics* 84: 323–29.

Walker, T. D., and J. W. Valentine. 1984. Equilibrium Models of Evolutionary Species Diversity and the Number of Empty Niches. *American Naturalist* 124: 887–99.

Walsh, B., and M. W. Blows. 2009. Abundant Genetic Variation + Strong Selection = Multivariate Genetic Constraints: A Geometric View of Adaptation. *Annual Review of Ecology, Evolution, and Systematics* 40: 41–59.

Wang, L., B. Spira, Z. Zhou, L. Feng, R. P. Maharjan, X. Li, et al. 2010. Divergence Involving Global Regulatory Gene Mutations in an *Escherichia coli* Population Evolving Under Phosphate Limitation. *Genome Biology and Evolution* 2: 478–87.

Weaver, S. C., A. C. Brault, W. Kang, and J. J. Holland. 1999. Genetic and Fitness Changes Accompanying Adaptation of an Arbovirus to Vertebrate and Invertebrate Cells. *Journal of Virology* 73: 4316–26.

Weinreich, D. M., N. F. Delaney, M. A. Depristo, and D. L. Hartl. 2006. Darwinian Evolution Can Follow Only Very Few Mutational Paths to Fitter Proteins. *Science* 312: 111–14.

Wellings, C. R. 2007. *Puccinia striiformis* in Australia: A Review of the Incursion, Evolution, and Adaptation of Stripe Rust in the Period 1979–2006. *Australian Journal of Agricultural Research* 58: 567–75.

Wellings, C. R., and R. A. Mcintosh. 1987. *Puccinia striiformis* f. sp. *tritiei* in Eastern Australia— Possible Means of Entry and Implications for Plant Quarantine. *Plant Pathology* 36: 239–41.

Whitlock, M. C. 1996. The Red Queen Beats the Jack-of-All-Trades: The Limitations on the Evolution of Phenotypic Plasticity and Niche Breadth. *American Naturalist* 148: S65–77.

Wichman, H. A., M. R. Badgett, L. A. Scott, C. M. Boulianne, and J. J. Bull. 1999. Different Trajectories of Parallel Evolution during Viral Adaptation. *Science* 285: 422–24.

Wichman, H. A., and C. J. Brown. 2010. Experimental Evolution of Viruses: Microviridae as a Model System. *Philosophical Transactions of the Royal Society B: Biological Sciences* 365: 2495–501.

Wichman, H. A., J. Millstein, and J. J. Bull. 2005. Adaptive Molecular Evolution for 13,000 Phage Generations: A Possible Arms Race. *Genetics* 170: 19–31.

Wichman, H. A., L. A. Scott, C. D. Yarber, and J. J. Bull. 2000. Experimental Evolution Recapitulates Natural Evolution. *Philosophical Transactions of the Royal Society B: Biological Sciences* 355: 1677–84.

Williams, G. C. 1966. *Adaptation and Natural Selection.* Princeton, NJ: Princeton University Press.

Withler, F. C. 1982. Transplanting Pacific Salmon. *Canadian Technical Report of Fisheries and Aquatic Sciences* 1079.

Wong, A., and R. Kassen. 2011. Parallel Evolution and Local Differentiation in Quinolone Resistance in *Pseudomonas aeruginosa*. *Microbiology* 157: 937–44.

Wong, A., N. Rodrigue, and R. Kassen. 2012. Genomics of Adaptation during Experimental Evolution of the Opportunistic Pathogen *Pseudomonas aeruginosa*. *PLoS Genetics* 8: e1002928.

Woods, R. J., J. E. Barrick, T. F. Cooper, U. Shrestha, M. R. Kauth, and R. E. Lenski. 2011. Second-Order Selection for Evolvability in a Large *Escherichia coli* Population. *Science* 331: 1433–36.

Woods, R., D. Schneider, C. L. Winkworth, M. A. Riley, and R. E. Lenski. 2006. Tests of Parallel Molecular Evolution in a Long-Term Experiment with *Escherichia coli*. *Proceedings of the National Academy of Sciences USA* 103: 9107–12.

Wright, S. 1982. The Shifting Balance Theory and Macroevolution. *Annual Review of Genetics* 16: 1–19.

Yang, L., L. Jelsbak, R. L. Marvig, S. Damkiær, C. T. Workman, M. H. Rau, et al. 2011. Evolutionary Dynamics of Bacteria in a Human Host Environment. *Proceedings of the National Academy of Sciences USA* 108: 7481–86.

Yeaman, S., and M. C. Whitlock. 2011. The Genetic Architecture of Adaptation Under Migration-Selection Balance. *Evolution* 65: 1897–911.

Yoder, J. B., E. Clancey, S. Des Roches, J. M. Eastman, L. Gentry, W. Godsoe, et al. 2010. Ecological Opportunity and the Origin of Adaptive Radiations. *Journal of Evolutionary Biology* 23: 1581–96.

Zambrano, M. M., D. A. Siegele, M. Almirón, A. Tormo, and R. Kolter. 1993. Microbial Competition: *Escherichia coli* Mutants That Take Over Stationary Phase Cultures. *Science* 259: 1757–60.

Zeyl, C. 2005. The Number of Mutations Selected during Adaptation in a Laboratory Population of *Saccharomyces cerevisiae*. *Genetics* 169: 1825–31.

Zeyl, C., T. Vanderford, and M. Carter. 2003. An Evolutionary Advantage of Haploidy in Large Yeast Populations. *Science* 299: 555–58.

Zhang, Q.-G., R. J. Ellis, and H. C. J. Godfray. 2012. The Effect of a Competitor on a Model Adaptive Radiation. *Evolution* 66: 1985–90.

Zhong, S., A. Khodursky, D. E. Dykhuizen, and A. M. Dean. 2004. Evolutionary Genomics of Ecological Specialization. *Proceedings of the National Academy of Sciences USA* 101: 11719–24.

Zhong, S., S. P. Miller, D. E. Dykhuizen, and A. M. Dean. 2009. Transcription, Translation, and the Evolution of Specialists and Generalists. *Molecular Biology and Evolution* 26: 2661–78.

Zinser, E. R., and R. Kolter. 1999. Mutations Enhancing Amino Acid Catabolism Confer a Growth Advantage in Stationary Phase. *Journal of Bacteriology* 181: 5800–5807.

Index